Palgrave Studies in Impact Finance

Series Editor
Mario La Torre
Sapienza University of Rome
Rome, Italy

The *Palgrave Studies in Impact Finance* series provides a valuable scientific 'hub' for researchers, professionals and policy makers involved in Impact finance and related topics. It includes studies in the social, political, environmental and ethical impact of finance, exploring all aspects of impact finance and socially responsible investment, including policy issues, financial instruments, markets and clients, standards, regulations and financial management, with a particular focus on impact investments and microfinance.

Titles feature the most recent empirical analysis with a theoretical approach, including up to date and innovative studies that cover issues which impact finance and society globally.

More information about this series at
http://www.palgrave.com/gp/series/14621

Marco Migliorelli · Philippe Dessertine
Editors

Sustainability and Financial Risks

The Impact of Climate Change, Environmental Degradation and Social Inequality on Financial Markets

Editors
Marco Migliorelli
IAE Université Paris 1
Panthéon-Sorbonne
(Sorbonne Business School)
Paris, France

European Commission
Brussels, Belgium

Philippe Dessertine
IAE Université Paris 1
Panthéon-Sorbonne
(Sorbonne Business School)
Paris, France

ISSN 2662-5105 ISSN 2662-5113 (electronic)
Palgrave Studies in Impact Finance
ISBN 978-3-030-54529-1 ISBN 978-3-030-54530-7 (eBook)
https://doi.org/10.1007/978-3-030-54530-7

This Palgrave Macmillan imprint is published by the registered company Springer Nature Switzerland AG
The registered company address is: Gewerbestrasse 11, 6330 Cham, Switzerland

To Maël and Amélie

PREFACE

The issue of the sustainability of the human activities is increasingly growing in importance in the agendas of both governments and international organisations. Today, the fight against climate change, the preservation of the environment and the ecosystems, the reduction of inequalities are constantly at the heart of the political and societal debate. The Paris Agreement and the Sustainable Development Goals (SDG) are in this respect the cornerstones of the engagement of the international community towards a climate-neutral economy and a fairer society. In such a context, the role of finance has been consolidating as a key enabling factors to reach the most ambitious policy objectives. Frameworks such as sustainable finance, green finance, climate finance, among others, have progressively emerged and are getting more and more support from policymakers in the attempt to mainstream such new way of doing finance.

Nevertheless, it can be observed that the relationship between finance and sustainability has been thus far analysed mainly in one direction, that is focussing on the key qualifying role of finance in reaching climate, environmental and social goals. On the contrary, little attention has been given so far to the assessment of the possible impacts of sustainability-related risks on financial actors and markets. That is, on how factors such as climate change, environmental degradation or social inequality, among others, may trigger new financial risks. This nexus is only recently emerging as a potential source of concerns for both the financial industry and policymakers. However, its relevance is expected to grow significantly in the near future along with the consolidation of some of the sustainability-related risks (e.g. climate change).

Overlooking the importance of the impact of sustainability-related risks on financial actors may have two main drawbacks. On the one side, it can result in a poor assessment of the commitment of the financial industry to support the different types of sustainable investments, in particular in the geographical areas and in the economic sectors expected to be more affected by sustainability-related risks. Situations in which banks or insurance companies may refuse to take-in additional financial risk when highly dependent of sustainable-related risks can indeed materialise. This would be for example the case of insurance companies refusing to insure households living in areas subject to increasing risk of floods, or of banks limiting credit to farmers in regions hit by increasing desertification. On the other side, the possible underestimation by banks and insurance companies of the long-term impact of sustainability-related risks on their businesses could eventually harm financial stability in the case this underestimation becomes systematic and widespread. As a matter of fact, very little evidence is today available as concerns the financial actors' assets under the risk of climate change and other sustainability-related risks, as well as on the possible strategies to integrated sustainability-related risks in existing risk management frameworks and pricing schemes.

This edited book has the objective to deepen the existing scarce knowledge on the relationship between sustainability-related risks and financial risks. To do that, it is structured in five chapters. In Chapter 1, Marco Migliorelli presents an overview of the key characteristics of the nexus between sustainability and financial risks. To this extent, the chapter first describes the role of finance in reaching a more sustainable society, in particular in the context of the Paris Agreement and the Sustainable Development Goals (SDG). Then, it identifies the key areas of financial risks stemming from sustainability-related risks, in particular as concerns climate change, environmental degradation, social inequality, policy and technology shifts. Finally, it explores the possible negative consequence of the full consideration of sustainability-related risks as source of financial risks, both in terms of pricing adjustments and, potentially, new market failures. In Chapter 2, Giorgio Caselli and Catarina Figueira deepen the analysis on the possible impacts of climate change on the banking and insurance industries. To this purpose, the chapter presents the main channels through which the physical, transition and liability risks of climate change might translate into financial risks for banks and insurance companies, along with the key data available to date. Chapter 3, written by Olaf Weber, Truzaar Dordi and Adeboye Oyegunle, is specifically dedicated to the issue of stranded assets in a society in transition towards a low-carbon economy. To this extent, the chapter first gives a historical account of stranded assets and presents a systematic review of the current state of the literature on the subject. Then, it proposes a comprehensive approach to understanding the multitude of factors resulting in stranded asset risk, by also including case studies to show how responses to stranded asset risks vary by region. Finally, it offers a research agenda for future studies, addressing some of the limitations to current research. In Chapter 4, Marco Migliorelli and Vladimiro Marini give an overview of the main strategic and organisational implications for financial institutions when fully considering the actual and potential impacts of sustainability-related risks on their businesses. In this respect, the chapter first analyses the main elements of the risk management framework that require a specific development, in particular within the perimeter of competence of the management board, the risk management function and the operational business units (these latter being the "first line of defence"). Then, the chapter discusses the issue of disclosing sustainability-related information, by illustrating existing standards and assessing the effectiveness of market discipline in the actual policy context. Finally, in Chapter 5, Marco Migliorelli, Nicola Ciampoli and Philippe Dessertine discuss the possible impact of the sustainability-related risks on financial stability. To this extent, they first identify the areas in which evolution in the practices of financial intermediaries are due to better manage sustainability-related risks. Hence, they discuss a set of policy actions necessary to both mitigate and control for the potential impact of sustainability-related risk on financial risks, in this way safeguarding financial stability. In this respect, particular is given to the need of evolving the prudential supervisory approaches and to the leading role of central banks.

Paris, France   Marco Migliorelli
                                                                           Philippe Dessertine

Acknowledgments We would like to thank the contributors. This book would have not come to fruition without their strong expertise and exceptional commitment.

CONTENTS

Notes on Contributors

Giorgio Caselli is a research fellow at the Centre for Business Research of the University of Cambridge. He holds a Ph.D. in financial economics from Cranfield University. His research interests lie at the intersection of business economics and finance, with a specific focus on how firm behaviour shapes macroeconomic outcomes.

Nicola Ciampoli is an adjunct professor of financial intermediaries at the LUMSA University in Rome and an independent advisor and certified accountant. He holds a Ph.D. from the University of Rome Tor Vergata. His research interests include sustainable value in banking and banking regulation.

Philippe Dessertine is a full professor of finance at the IAE of the University Paris 1 Panthéon-Sorbonne (Sorbonne Business School). He is also the director of the *Institut de Haute Finance*, Paris, France and a former member of the *Haut Conseil des Finances Publiques* in France. He is author of several publications on the role of finance in modern society.

Truzaar Dordi is a Ph.D. candidate in sustainability management at the University of Waterloo, from which he also holds a Master in sustainability management. His research interests cover the fields of climate finance, energy policy and risk management, with particular emphasis on financial system stability along a low-carbon transition.

Catarina Figueira is a professor of applied economics and policy and head of the Economics and Finance Centre at Cranfield University. She also leads the Economic Policy and Performance Group at Cranfield School of Management and has set up and directs the UK's first Mastership in retail and digital banking. Her research interests lie in the area of financial institutions' performance and the effects of regulatory changes.

Vladimiro Marini is a junior assistant professor at Roma Tre University. He holds a Ph.D. in banking and finance at the University of Rome Tor Vergata. His research activity is primarily focused on private equity investments, in particular as concerns skills and competences acquisition by fund managers.

Marco Migliorelli is an economist at the European Commission.[1] He also performs academic research at the IAE of the University Paris 1 Panthéon-Sorbonne (Sorbonne Business School) as associated researcher. He holds a Ph.D. in banking and finance from the University of Rome Tor Vergata. His main research interests include green finance and sustainable finance, cooperative banking, and financing instruments innovation.

Adeboye Oyegunle is a Ph.D. candidate in sustainability management at the University of Waterloo, from which he also holds a Master in sustainability management. His research interests include climate finance, climate scenarios and risk management, with a focus on financial sustainability regulations and guidelines.

Olaf Weber is a full professor at University of Waterloo, where he also holds the Chair in Sustainable Finance, and the editor in chief of the *Journal of Sustainable Finance and Investment*. He holds a Ph.D. from the University of Bielefeld. His research interests lay in the connection between financial sector players and sustainable development, and the link between sustainability and financial performance of enterprises.

---

[1] The contents included in this book do not necessarily reflect the official opinion of the European Commission. Responsibility for the information and views expressed in the book lies entirely with the authors.

LIST OF FIGURES

LIST OF TABLES

CHAPTER 1

The Sustainability–Financial Risk Nexus

Marco Migliorelli

Abstract This chapter gives an overview of the relationship nowadays linking sustainability-related risks (stemming from climate change, environmental degradation, social inequality, policy and technology shifts) and financial risks. Two main conclusions highlight the importance of this nexus. First, the expected consolidation of sustainability-related risks in the near future has the potential to produce a widespread impact on the financial results of both banks and insurance companies. Second, the full consideration by financial actors of sustainability-related risks may lead in some geographical areas and for some economic sectors to significant pricing adjustments and to new market failures (in terms of credit cutbacks and non-insurability of risks). The chapter concludes

M. Migliorelli (✉)
IAE Université Paris 1 Panthéon-Sorbonne (Sorbonne Business School), Paris, France
e-mail: Marco.Migliorelli@ec.europa.eu

European Commission, Brussels, Belgium

© The Author(s) 2020   1
M. Migliorelli and P. Dessertine (eds.), *Sustainability and Financial Risks*, Palgrave Studies in Impact Finance,
https://doi.org/10.1007/978-3-030-54530-7_1

by proposing a structured taxonomy systematically linking sustainability-related risks and financial risks.

Keywords Sustainability-related risks · Climate change-related risks · Physical risk · Transition risk · Financial risks · Sustainable finance

1.1   INTRODUCTION

When it comes to the discussion on the relation between sustainability and finance, the policy and academic debate has been focused thus far on the possible role of the latter to support the transition towards a climate-neutral economy and a fairer society. In this respect, concepts and frameworks such as sustainable finance, green finance or climate finance have progressively emerged.[1] These concepts and frameworks have also been consolidating within the financial industry in the form of new financial instruments (e.g. green bonds and sustainable funds), listing options (e.g. dedicated segments for sustainable securities in several stock exchanges worldwide), certification possibilities (e.g. green and climate labels for financial securities) or specific financing supporting initiatives (e.g. the World Bank or the European Investment Bank sustainability programmes). A new stream of literature is also progressively emerging dealing with these matters (e.g. Lehner 2016; Ziolo and Sergi 2019; Migliorelli and Dessertine 2019a).

Nevertheless, little attention has been given so far to the specific relationship linking sustainability and financial risks. That is, to the discussion on how factors such as climate change, environment degradation or social inequality, among others, can impact financial actors and markets. Indeed, this relationship, which is referred here as the *sustainability-financial risk nexus*, is of the utmost importance and may have systemic-wise consequences. For some observers, the failure of the various components of the financial industry to correctly integrate sustainability-related risks into financial risks frameworks may represent in the longer term a threat to the stability of the financial system as a whole (e.g. EC 2018a; BIS 2020).

---

[1] See for example UNEP (2016) or Berrou et al. (2019a).

To deepen the analysis on this issue, this chapter is structured as follows. First, Sect. 1.2 gives an overview of the role played nowadays by finance in fostering sustainability. To do that, the political and societal processes culminated with the adoption of the Sustainable Development Goals (SDG) and the signature of the Paris Agreement in 2015 are presented, as well as the expected contribution of finance in the resulting agendas. Then, Sect. 1.3 introduces the role of the *sustainability-financial risk nexus* within the general sustainable finance framework. In this respect, a review of the (scarce) literature dealing with this issue is also given. This dissertation is followed by Sect. 1.4, proposing a more comprehensive approach to the understanding of the relation between sustainability-related risks and financial risks. To this extent, a structured taxonomy linking the different types of risks is proposed. Finally, Sect. 1.5 concludes with a scrutiny of the key element of pricing of financial services when fully considering sustainability-related risks. Such an analysis includes the recognition of possible new market failures resulting from the progressive consolidation of these risks.

1.2 THE ROLE OF FINANCE IN FOSTERING SUSTAINABILITY

1.2.1 *The Sustainable Development Goals (SDG) and the Paris Agreement*

The concern of the sustainability of human activities have been discussed for decades (e.g. Renneboog et al. 2008; Berrou et al. 2019b). However, a significant acceleration in the political and societal debate has been observed only in the last few years. In this respect, the adoption of the Sustainable Development Goals (SDG) in September 2015 and the Paris Agreement[2] reached in December of the same year landmarked a new

---

[2] The Paris Agreement resulted from the United Nations Framework Convention on Climate Change (UNFCCC), an international environmental treaty that aims to limit global greenhouse gas (GHG) emissions and that is still in force today. Starting from 1995, signatories of the UNFCCC met on a yearly basis, through the Conferences of the Parties (COP). In 1997, as result of the conference held in Kyoto (COP 3), the Kyoto Protocol extended on the UNFCCC and led to the establishment of the first global legally binding obligation addressing climate change. The Paris Agreement was signed during the COP 21.

era for the fight against climate change and the transition towards a sustainable economy.[3]

The SDG are part of the "2030 Agenda for Sustainable Development" adopted by the United Nations (UN) General Assembly. The Agenda is "a plan of action for people, planet and prosperity. It also seeks to strengthen universal peace in larger freedom" (UN 2015). The SDG, to be achieved by 2030, have the merit to clearly identify the priorities of the international community in the attempt to reach a sustainable society, highlighting the importance of protecting the environment, of ensuring decent living conditions for all human beings and limiting the negative impacts of economic development. Table 1.1 reports the 17 SDG. In addition, 169 targets and 242 global indicators were also set to monitor the progress towards the realisation of the goals. In point of fact, the SDG reflect all the three distinctive dimensions of sustainable development: the economic, social and ecological dimensions. The wide acceptance of the SDG at the highest political levels represented with no doubt an important success and a significant step forward for the recognition of sustainability as one of key issues to be solved in the interest of humankind as whole.

Resulting from a parallel process, the Paris Agreement was conceived within the United Nations Framework Convention on Climate Change (UNFCCC), a global environmental treaty aiming at limiting global greenhouse gas (GHG) emissions. Starting from 1995, signatories of the UNFCCC have met on a yearly basis, through the Conferences of the Parties (COP). The Paris Agreement was signed during the COP 21,[4] when world leaders committed to strengthen the global response to the threat of climate change by "holding the increase in the global average temperature to well below 2°C above pre-industrial levels and pursuing efforts to limit the temperature increase to 1.5°C". To reach these ambitious objectives, appropriate mobilisation and provision of financial resources, a new technology framework and enhanced capacity-building were given specific and unprecedented attention. The agreement

---

[3] Among the other noteworthy initiatives on the defence of the environment, in May 2015 the Pope Francis addressed the subject of environmental degradation and climate change in a historical encyclical letter "Laudato sí" on "Care for Your Common Home".

[4] The UNFCCC had some encouraging results already before COP 21. In 1997, as result of the conference held in Kyoto (COP 3), the Kyoto Protocol led to the establishment of the first global legally binding obligation addressing climate change.

Table 1.1 Sustainable Development Goals (SDG)

#	Sustainable Development Goal	Short description
SDG 1	No poverty	End poverty in all its forms everywhere
SDG 2	Zero hunger	End hunger, achieve food security and improved nutrition and promote sustainable agriculture
SDG 3	Good health and well-being	Ensure healthy lives and promote well-being for all at all ages
SDG 4	Quality education	Ensure inclusive and equitable quality education and promote lifelong learning opportunities for all
SDG 5	Gender equality	Achieve gender equality and empower all women and girls
SDG 6	Clean water and sanitation	Ensure availability and sustainable management of water and sanitation for all
SDG 7	Affordable and clean energy	Ensure access to affordable, reliable, sustainable and modern energy for all
SDG 8	Decent work and economic growth	Promote sustained, inclusive and sustainable economic growth, full and productive employment and decent work for all
SDG 9	Industry, innovation and infrastructure	Build resilient infrastructure, promote inclusive and sustainable industrialisation, and foster innovation
SDG 10	Reduced inequalities	Reduce income inequality within and among countries
SDG 11	Sustainable cities and communities	Make cities and human settlements inclusive, safe, resilient and sustainable
SDG 12	Responsible consumption and production	Ensure sustainable consumption and production patterns
SDG 13	Climate action	Take urgent action to combat climate change and its impacts by regulating emissions and promoting developments in renewable energy

(continued)

Table 1.1 (continued)

#	Sustainable Development Goal	Short description
SDG 14	Life below water	Conserve and sustainably use the oceans, seas and marine resources for sustainable development
SDG 15	Life on land	Protect, restore and promote sustainable use of terrestrial ecosystem, sustainably manage forests, combat desertification, and halt and reverse land degradation and halt biodiversity loss
SDG 16	Peace, justice and strong institutions	Promote peaceful and inclusive societies for sustainable development, provide access to justice for all and build effective, accountable and inclusive institutions at all levels
SDG 17	Partnerships for the goals	Strengthen the means of implementation and revitalise the global partnership for sustainable development

Source Author's elaboration based on the SDG description as given in the 2030 Agenda for Sustainable Development (UN 2015)

requires all Parties to put forward their efforts through Nationally Determined Contributions (NDC) and to report regularly on their emissions and on their implementation efforts. The Parties also bore a responsibility to meet every five years on the subject and set up a robust, transparent and accountable reporting system to track their progresses.[5] Although the global reach of the Paris Agreement is undeniable, further work is still needed to ensure its concrete impact on climate change (Berrou et al. 2019b). In fact, the agreement is only partially legally binding and there are no means of systematically verifying if the Parties are reaching their objectives.[6] Some important items were also discarded from the

[5] The objectives that were announced during the agreements will be revised in 2020, and once every five years after that initial revision. An overall assessment will be performed in 2023, and, once more, will occur every five years.

[6] In addition, in June 2017, United States President Donald Trump announced his intention to withdraw his country from the Paris Agreement. Under the agreement itself, the earliest effective date of withdrawal for the United States is November 2020.

debate, including carbon pricing and the possible discontinuation of fossil fuel extractions. Furthermore, in 2018, the Intergovernmental Panel on Climate Change (IPCC)—the United Nations body for assessing the science related to climate change—launched the alarm stating that the world needs to limit temperature increase to 1.5 °C with respect to pre-industrial levels to reduce the likelihood of extreme weather events and emphasised that GHG emissions need to be reduced with far more urgency than previously assumed (IPCC 2018).[7]

The adoption of the SDG and the Paris Agreement and the growing awareness of the civil society for sustainability issues are progressively imposing a new agenda to both governments and international institutions (e.g. EC 2019a). The changeover implies a deep reflection on the economic and social structures today in place and needs strong political commitment, ambitious technology investments, adapted regulations and likely a change in the consumption and behavioural patterns of the population (e.g. EC 2018b). In such a context, the availability of financial resources to support the transition has consolidated as an essential enabling factor.[8]

1.2.2   The Rise of Sustainable Finance

Defining precisely what it is today called sustainable finance is not an easy task. As a matter of fact, financial institutions, governments and international organisations tend to create definitions according to their underlying motivations (UNEP 2016; IFC 2017). In addition, trough time a number of possibilities to account for the connection between finance and sustainability have flourished. Among them, it should be highlighted the concern with environmental, social and corporate governance (ESG) criteria (e.g. Friede et al. 2015), the impact investing and the social responsible investing (SRI) approaches (e.g. Vandekerckhove et al. 2012; Hebb 2013), the analysis of the impact of financial development on environment degradation (e.g. Tamazian et al. 2009), the concern with climate change and human rights (e.g. Alm and Sievänen 2013), the

---

[7] In particular, net-zero carbon emissions at global level need to be achieved not beyond the half of this century and neutrality for all other GHG not much later.

[8] As an example, investments of around EUR520–575 billion annually have been estimated to be necessary in the EU only in order to achieve a net-zero GHG economy in the 2050 horizon (*Source* EC 2018b).

assessment of the effect of finance in terms of negative externalities (e.g. Ziolo et al. 2019), the new role of sustainable finance for financial institutions having a formal dual bottom-line approach and for which financial performance need to coexist with social goals (e.g. Migliorelli 2018).

Nevertheless, the framework provided by the SDG can today be used as a new reference in the attempt to better circumscribe the perimeter of action of sustainable finance and its various components. In this respect, sustainable finance may be considered to embrace all the financial stocks and flows mobilised to achieve the SDG, irrespectively of their labelling or the technical implementation of the underlying financial instruments. Furthermore, what today is generally refereed to green finance and climate finance can be considered to be specific parts of the wider sustainable finance landscape.[9] To this extent, green finance can be referred to the financial stocks and flows aiming at supporting the achievement of the environment-related SDG,[10] while climate finance can be associated to that component of green finance focussing on climate action (in the form of climate change mitigation and climate change adaptation[11]). These relations are graphically reported in Fig. 1.1.

The various components of sustainable finance have experienced a remarkable growth in recent years, and in particular as it concerns green finance. For example, from the first issuance by the European Investment Bank in 2007, the market of green bonds has registered average annual two-digit growth, with new emissions being over USD160 billon in 2018,[12] while sustainable or green equivalents of traditional securities are today getting available for the different types of investors. In point of fact, a large part of the financial industry and several policymakers

[9] For a wider dissertation on green finance and the challenges it faces, see Migliorelli and Dessertine (2019a).

[10] For a discussion on the definition of green finance, see Berrou et al. (2019a).

[11] Climate change mitigation usually refers to efforts to reduce or prevent emission of GHG. Climate change adaptation normally concerns the adjustments in ecological, social or economic systems in response to actual or expected climatic modifications and their effects or impacts.

[12] See Berrou et al. (2019b).

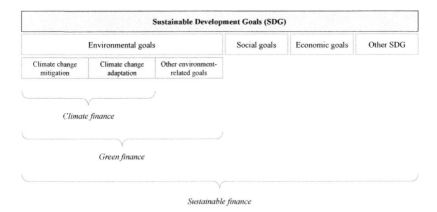

Fig. 1.1 SDG, sustainable finance and its components (*Source* Adapted from UNEP 2016)

have embraced the change and are putting in place a number of initiatives in the attempt to mainstream sustainable finance.[13,14] The European Commission's "Action Plan for financing sustainable growth" issued in 2018 and its follow-up initiatives is probably the most noteworthy example of this commitment (EC 2018a).

[13] The growth of sustainable finance in the last decade should be also related to a strong commitment of the major stock exchanges worldwide. Financial centres such as London, Paris, Luxembourg, Copenhagen, Amsterdam in Europe, Shanghai and Beijing in China, San Francisco and Los Angeles in the United States, Vancouver and Montreal in Canada have taken the lead and are progressively improving the quality and depth of their sustainable finance offer. To this extent, dedicated listings for sustainable finance and green finance securities have emerged.

[14] Nevertheless, some challenges still exist and mainstreaming sustainable finance can be considered a long-term objective. In particular, clearly identifying the sectors or activities eligible for sustainable finance, better assessing the (still unclear) financial benefits for issuers of sustainable securities, coping with the lack of incentives for market actors of entirely factoring in the sustainability-related risks in their investment decisions are some of these challenges. In addition, to effectively mainstream sustainable finance, some conditions need to be fulfilled. Namely, environmental risks are properly included in the investors' decision-making processes, market demand is effectively channelled towards sustainable investments, additionality is adequately encouraged by policymakers, the banking sector is fully engaged in the transition. For a wider discussion on these subjects, with a focus on green finance, see also Migliorelli and Dessertine (2019b).

1.3   SUSTAINABILITY AND FINANCE: A TWO-WAY RELATIONSHIP

1.3.1   *Positioning the* Sustainability-Financial Risk Nexus

The policy and academic debate are consolidating around the analysis of how finance can contribute to the transition towards a sustainable society by ensuring that the necessary financial resources are available when and where needed. In this respect, the role of banks and other financial actors is a key one when simply considering their traditional function of funds intermediaries. As a matter of fact, adequate financing to the sustainability transition cannot be achieved without the full involvement of the financial industry. However, little attention has been given thus far to the possible impact of sustainability-related risks on financial actors, that is to the *sustainability-financial risk nexus*. Factors such as climate change, environmental degradation or social inequality, and others, can indeed result in direct or indirect financial risks for financial actors. An example can help illustrating this issue. Considering climate change, abounded evidence exists today demonstrating that the continuous increase in GHG emissions in the atmosphere ultimately results in a substantial increase in the frequency and magnitude of climate change-related extreme weather events such as droughts, floods or storms (e.g. IPCC 2018). Beyond the (regrettable) direct consequences on the populations and their social implications, extreme weather events may also have relevant impacts on insurance companies and banks, as unexpected and important reductions in the productivity of the economic assets typically materialise in the areas affected. For insurance companies, this can produce unexpected higher levels of payments on the previously insured risks. For banks, higher levels of impairments on outstanding credits due to higher rates of insolvency of their clients.

Underestimating the impact of sustainability-related risks on financial actors may have two main drawbacks. Firstly, it can result in a flowed assessment of the commitment and the capacity of the financial industry to support sustainable investments, in particular in the areas expected to be more affected by sustainability-related risks. Situations in which banks or insurance companies refuse to take-in additional financial risk when highly dependent of sustainable-related risks can eventually materialise. This would be for example the case of banks limiting credit to farmers in regions hit by increasing desertification, as considered to be less productive in the mid-term. Or of insurance companies refusing to insure households living in areas subject to increasing risk of floods. On the other side, a systematic underestimation by banks and insurance companies of the long-term impact of sustainability-related risks on their core businesses could bring to a situation in which financial stability[15] can be undermined. As little historical data (and knowledge) is today available for financial actors as concerns the possible incidence of sustainability-related risks, and the occurrence of these risks is expected to grow in future both in terms of frequency and magnitude (so that past experience cannot be used to predict the future), eventually very little information is today available as concerns the financial actors' assets under the risk of climate change or other sustainability-related risks (e.g. ECB 2019).[16]

Hence, the *sustainability-financial risk nexus* merits today a throughout attention and it should be considered as a crucial element of the sustainable finance framework.[17] Clearly, solving the *sustainability-financial risk nexus* implies two separate dimensions of analysis. On the one hand, the assessment of the possibilities of reducing the magnitude of sustainability-related risks. This can be done through policy and societal actions aiming at fostering a climate-neutral economy and a fairer society.

---

[15] Financial stability can be defined as a condition in which the financial system— which comprises financial intermediaries, markets and market infrastructures—is capable of withstanding shocks and the unravelling of financial imbalances. This mitigates the likelihood of disruptions in the financial intermediation process that are systemic, that is, severe enough to trigger a material contraction of real economic activity (ECB website, consultable here: https://www.ecb.europa.eu/pub/financial-stability/fsr/html/ecb.fsr201 911~facad0251f.en.html#toc1).

[16] Based on EC (2019b), weather-related disasters caused a record EUR 283 billion in economic damages in 2017 and could affect up to two-thirds of the European population by 2100 compared with 5% today.

[17] In this respect, a noteworthy initiative is the establishment of the Network for Greening the Financial System (NGFS), launched at the One Planet Summit in Paris in December 2017 under the initiative of the Banque de France. Composed by more than 30 central banks and supervisory bodies (including Banco de España, Bank of England, Bank of Finland, Banque Centrale du Luxembourg, Deutsche Bundesbank, European Banking Authority, European Central Bank, Japan FSA, National Bank of Belgium, Oesterreichische National Bank, the People's Bank of China, the Reserve Bank of Australia, Reserve Bank of New Zealand), it aims on a voluntary basis to exchange experiences and best practices, to contribute to the development of environment and climate risk management in the financial sector, and to mobilise mainstream finance to support the transition towards a sustainable economy. In 2019, the NGFS issued the first comprehensive report on climate change as source of financial risk (NGFS 2019).

As a matter of fact, the development of sustainable finance securities, products and services can be embedded in this dimension. On the other hand, the consideration of the *sustainability-financial risks nexus* triggers the need of controlling for the impact of the key sustainability-related risks on financial actors. In this respect, an assessment of the existing risk management frameworks should be systematically carried out to test for their capacity to take into account these new risks.

1.3.2   Sustainability-Related Risks and Observed Channels of Transmission to the Financial Markets

Limited literature exists dealing with the *sustainability-financial risk nexus*. In this section the main references to date are reported as concerns the impact of climate change (in the form of physical risk, transition risk and liability risk), distressed commodity markets, environmental degradation and social inequality.[18]

1.3.2.1   Climate Change: Physical Risk, Transition Risk and Liability Risk

Central banks have been among the first actors to recognise that even though significant macroeconomic effects from climate change may occur in a somehow distant future, some impacts are already beginning to be felt (ECB 2019). As a consequence, in the last few years, and in line with their activity of supervision and control of systemic risks, they have started to identify some specific financial risks linked to climate change (BoE 2015; TCFD 2017; ACPR 2019; ECB 2019). Namely:

- Physical risks, defined as the impacts today on insurance liabilities and the value of financial assets that arise from climate and weather-related events that may damage property or disrupt trade.[19]

---

[18] Even if not linked to financial risks, sustainability-related risks have nevertheless recently started to be considered as crucial factors in the development of modern society. Extreme weather events, failure of managing climate change mitigation and adaptation, natural disasters, man-made environmental disasters, large-scale involuntary migration, biodiversity loss and ecosystem collapse, water crises, occupy seven positions in a top ten of risks by likelihood by the World Economic Forum (WEF 2019).

[19] The United Nations Environmental Programme Finance Initiative (UNEP FI) provides a methodology for assessing physical risk (UNEP FI 2018). It recommends

- Transition risks, that is the financial risks that could result from the process of adjustment towards a low-carbon economy, such as changes in policy, technology and physical risks that could prompt a reassessment of the value of a large range of assets as costs and opportunities become apparent (the case of stranded assets).
- Liability risks, that is the impacts that could arise tomorrow if parties who have suffered losses or damages from the effects of climate change seek compensation from those they hold responsible (such claims could come decades in the future, but have the potential to hit carbon extractors and emitters and, if they have liability cover, their insurers).

Nevertheless, the limitation to climate change and a substantial lack of data to properly assess the impact of these risks make this recognition still a marginal improvement in the understanding the relationship between sustainability-related risks and financial risks.[20]

considering both changes in average weather conditions and the more frequent occurrence of extreme events. To implement these exercises, it would be necessary to improve the available data, in particular on the geographical location of borrowers, to improve macroeconomic models that integrate the impact of climate change and to anticipate difficulties that the insurance sector could experience.

[20] Some first structured attempts to specifically analyse the incidence of these risks has been indeed made in Europe by the British Prudential Regulation Authority (PRA) in 2018 and by the French Autorité de contrôle prudentiel et de resolution (ACPR) in 2019. The PRA surveyed a number of UK banks on the possible incidence of climate change-related risks (PRA 2018). Relevant conclusions included: (i) for banks, the financial risks from climate change have tended to be beyond their planning horizons (for 90% of the UK banking sector these horizons averaged four years—before risks would be expected to be fully realised and prior to stringent climate policies taking effect); (ii) the majority of banks are beginning to treat the risks from climate change like other financial risks rather than viewing them simply as a corporate social responsibility issue; such banks start to oversight the financial risks from climate change and assign the overall responsibilities for setting the strategy, targets and risk appetite relating to these risks (including at board level); (iii) banks have begun considering the most immediate physical risks to their business models and have started to assess exposures to transition risks where government policy is already pulling forward the adjustment (this latter includes exposures to carbon-intensive sectors, consumer loans secured on diesel vehicles, and buy-to-let lending given new energy efficiency requirements). Similarly, the ACPR surveyed the main French banking groups (ACPR 2019). The main conclusions stemming from the survey were: (i) banking groups appear to have relatively little exposure to physical risk on the basis of currently available scenarios and expected impacts are mainly concentrated in low-vulnerability geographical areas (nevertheless, the industry seems to be aware that

1.3.2.2   Distressed Commodity Markets

Rising temperatures and changing patterns of precipitation can be expected to have direct impacts in particular on agriculture and fisheries (e.g. ECB 2019), even though with uneven influence between the different regions worldwide. In this respect, some regions are already substantially affected by both global climate variations and commodity price fluctuations.[21] This is valid also when considering that the impact of changing weather conditions on commodities' production and yields are strongly dependent of technology availability and sophistication (Brown and Funk 2008).

Today, financialisation of commodity markets can be considered a structural trend. In this respect, it can also be argued that commodities-driven fund management have become a proper investment style for many institutional investors (e.g. Adams and Glück 2015). This means that, as those institutional investors continue to target their managed funds into commodities, spillovers effects between commodities markets and financial markets will probably increase. Hence, higher volatility in the commodity markets can be considered today a specific source of concerns for fund managers, including when triggered by climate change.

1.3.2.3   Environmental Degradation and Social Inequality

Abundant and substantially unanimous literature today exists demonstrating the detrimental effect on the environment of the traditional model of economic development, in particular due to resources depletion and negative externalities (e.g. Tamazian et al. 2009; IPCC 2018). Land degradation, land erosion, waters and air pollution, deforestation are among the most visible signs of this pattern. In this vein, the behaviour of companies in terms of environmental and social consideration has

---

the full risk is not necessarily and fully transferable to the insurance sector); (ii) achieved progress in the area of transition risks was the most significant as banking institutions consider themselves being more directly exposed to this risk (in the mid-term), even though this trend is unevenly distributed across banking groups (institutions underlined that the horizon for transition risk is much closer to the one underlying their strategic thinking); (iii) most of respondents consider not to be exposed to liability risk in a material manner, even though the number of litigations is increasing at the international level and institutions are encouraged to seize this topic.

[21] For example, it has been observed that warming in the Indian Ocean and an increasingly concentrate precipitations (as in the case of hurricanes) could reduce main-season precipitation across vast parts of the Americas, Africa and Asia (Brown and Funk 2008).

been mostly studied in the framework of the analysis of the relationship between environmental, social and governance (ESG) performances and economic and financial performances. The large majority of studies show positive relationship of ESG performances on economic and financial performances, with the impact appearing to be stable over time (e.g. Friede et al. 2015). Nevertheless, the aspect of how environmental degradation or social inequality can negatively impact economic development and eventually financial markets and actors have thus far not been explored in depth.

1.4   A Wider Look
at the *Sustainability-Financial Risk Nexus*

A more comprehensive approach to the study of the *sustainability-financial risk nexus* can be proposed. In this respect, Table 1.2 suggests a taxonomy linking sustainability-related factors and risks to the corresponding risks for business, banks and insurance companies.[22] The relationship portrayed are assumed and not backed by data. Nevertheless, such taxonomy can help identifying potential indirect and direct financial risks for banks and insurance companies stemming from sustainability-related factors. In this respect, indirect risks for financial intermediaries have to be considered the ones coming from the exposure to sustainability-related factors by the clients (businesses) they serve.

Four main sustainability-related factors are considered: climate change, environmental degradation, social inequality, policy and technology shifts. To these main factors, specific sustainability-related risks potentially affecting businesses and financial actors are linked. For example, to climate change are associated risks of increase in the frequency and magnitude of floods, droughts and storms, of distressed commodity markets, of permanent change in climate conditions, of increase in the level of seas and of accusation from citizens to polluting businesses to cause climate change. These sustainability-related risks can be associated to concrete risks for businesses (hence also indirectly triggering risks for banks and insurance companies). For instance, the increase in the frequency and magnitude of floods, droughts and storms can result for businesses in loss of production, in a reduction in assets' value or in the disruption in

---

[22] For a similar exercise, limited to climate change, see TCFD (2017, pp. 10 and 11).

Table 1.2 A taxonomy for sustainability-related risks and financial risks

Sustainability-related factor	Sustainability-related risk	Risk for businesses	Risk for banks	Risk for insurance companies
Climate change	Increase in the frequency and magnitude of floods, droughts and storms	Loss of production [operational risk-related]	Increase in clients' insolvency risk [credit risk-related]	Higher payments on insured risks [physical risk and liquidity risk-related]
		Reduction in assets' value [operational risk-related]	– Reduction in the value of guarantees [credit risk-related] – Increase in clients' insolvency risk [credit risk-related]	Higher payments on insured risks [physical risk and liquidity risk-related]
		Disruption in the supply chain or in the operations [operational risk-related]	Increase in clients' insolvency risk [credit risk-related]	–
			Disruption in the operational activities [operational risk-related]	–
			–	Disruption in the operational activities [operational risk-related]
	Distressed commodity markets	Higher costs of production and/or cost of hedging	Increase in clients' insolvency risk [credit risk-related]	–

Sustainability-related factor	Sustainability-related risk	Risk for businesses	Risk for banks	Risk for insurance companies
		–	Increase in the volatility of the value of the investment portfolios [market risk-related]	–
		–	–	Increase in the volatility of the value of the investment portfolios [market risk-related]
	Permanent change in climate conditions	Loss in assets and land productivity	– Increase in clients' insolvency risk [credit risk-related] – Reduction in the value of guarantees [credit risk-related]	–
	Increase in the level of seas	Emerging adaptation costs	Increase in clients' insolvency risk [credit risk-related]	–
		Permanent loss of assets and land	– Increase in clients' insolvency risk [credit risk-related] – Loss of the value of guarantees [credit risk-related]	Higher payments on insured risks [physical risk and liquidity risk-related]

(continued)

Table 1.2 (continued)

Sustainability-related factor	Sustainability-related risk	Risk for businesses	Risk for banks	Risk for insurance companies
	Accusation from citizens to polluting businesses to cause climate change	Reputational risk	– Reputational risk due to accusation of financing polluting businesses – Increase in clients' insolvency risk [credit risk-related]	–
		Loss of clients	Increase in clients' insolvency risk [credit risk-related]	–
Environmental degradation	Soil, air or waters pollution	Production disruption or interruption	Increase in clients' insolvency risk [credit risk-related]	–
	Deforestation	*Increase in other risks linked to climate change, environmental degradation, social inequality, policy and technology shifts*		
	Loss of biodiversity	*Increase in other risks linked to climate change, environmental degradation, social inequality, policy and technology shifts*		
	Increase in the frequency and magnitude of epidemics or pandemics	Reduction in market demand or stop of production following containment measures	– Increase in clients' insolvency risk [credit risk-related] – Increase in the volatility of the value of the investment portfolios [market risk-related]	Increase in the volatility of the value of the investment portfolios [market risk-related]

Sustainability-related factor	Sustainability-related risk	Risk for businesses	Risk for banks	Risk for insurance companies
Judiciary actions from affected population towards polluters		Compensation due to proven responsibility [liability risk]	– Increase in clients' insolvency risk [credit risk-related] – Compensation due to possible joint responsibility [liability risk]	Higher payments on insured risks [liquidity risk-related]
		Reputational risk	Increase in clients' insolvency risk [credit risk-related]	–
Accusation from citizens to polluting businesses to cause environmental degradation		Reputational risk	– Reputational risk due to accusation of financing polluting businesses – Increase in clients' insolvency risk [credit risk-related]	–
		Loss of clients	Increase in clients' insolvency risk [credit risk-related]	–

(continued)

Table 1.2 (continued)

Sustainability-related factor	Sustainability-related risk	Risk for businesses	Risk for banks	Risk for insurance companies
Social inequality	Unfair treatment of workers	Reputational risk	– Reputational risk due to accusation of financing unfair businesses practices – Increase in clients' insolvency risk [credit risk-related]	–
		Compensation due to proven responsibility [liability risk]	Increase in clients' insolvency risk [credit risk-related]	–
		–	Reputational risk Compensation due to proven responsibility [liability risk]	–
		–	–	Reputational risk Compensation due to proven responsibility [liability risk]
	Discriminatory treatment of women	Reputational risk	– Reputational risk due to accusation of financing unfair businesses practices – Increase in clients' insolvency risk [credit risk-related]	–

Sustainability-related factor	Sustainability-related risk	Risk for businesses	Risk for banks	Risk for insurance companies
		Compensation due to proven responsibility [liability risk] —	Increase in clients' insolvency risk [credit risk-related] Reputational risk Compensation due to proven responsibility [liability risk] —	— — Reputational risk Compensation due to proven responsibility [liability risk] —
Discriminatory treatment of minorities		Reputational risk Compensation due to proven responsibility [liability risk] —	– Reputational risk due to accusation of financing unfair businesses practices – Increase in clients' insolvency risk [credit risk-related Increase in clients' insolvency risk [credit risk-related] Reputational risk Compensation due to proven responsibility [liability risk]	— —

(continued)

Table 1.2 (continued)

Sustainability-related factor	Sustainability-related risk	Risk for businesses	Risk for banks	Risk for insurance companies
		—	—	Reputational risk Compensation due to proven responsibility [liability risk]
	Social dumping (employees treated differently in different jurisdictions to save on costs)	Reputational risk	– Reputational risk due to accusation of financing unfair businesses practices – Increase in clients' insolvency risk [credit risk-related] Reputational risk	–
Policy and technology shifts	Reduction in public financing/policy-driven higher cost of financing for polluting assets	Higher cost of financing	Increase in clients' insolvency risk [credit risk-related]	Reputational risk –
	More stringent regulatory requirements in order to promote sustainability	Stranded assets [transition risk]	– Increase in clients' insolvency risk [credit risk-related] – Loss of the value of guarantees [credit risk-related]	–

Sustainability-related factor	Sustainability-related risk	Risk for businesses	Risk for banks	Risk for insurance companies
		Obsolete business lines	– Increase in clients' insolvency risk [credit risk-related] – Loss of the value of guarantees [credit risk-related] Reduction in the value of the stocks of non-sustainable industries in the investment portfolios [market risk-related]	– – Reduction in the value of the stocks of non-sustainable industries in the investment portfolios [market risk-related]
Technology improvements causing obsolescence of polluting assets	Stranded assets [transition risk]	–	– Increase in clients' insolvency risk [credit risk-related] – Loss of the value of guarantees [credit risk-related] Reduction in the value of the stocks of non-sustainable industries in the investment portfolios [market risk-related]	–

(continued)

Table 1.2 (continued)

Sustainability-related factor	Sustainability-related risk	Risk for businesses	Risk for banks	Risk for insurance companies
		–	–	Reduction in the value of the stocks of non-sustainable industries in the investment portfolios [market risk-related]

Notes The table shows the effects of sustainability-related risk events on the different economic actors. In particular, it links the risks for businesses with the consequent risks for banks and insurance companies. Nevertheless, some of the sustainability-related risk events can also impact directly banks and insurance companies. This is in particular the case for distressed commodity markets due to climate change, for social inequality-related risks and policy and technology shifts. Blank cells mean no risk for that actor (and hence no relation with the risks for the other actors shown in same line)
Source Author's elaboration

the supply chain or in the operations. These risks for businesses can be hence analysed with respect to corresponding risks for banks and insurance companies. To this extent, the reduction in assets' value of their clients can cause, for banks, a reduction in the value of the real guarantees (e.g. covering a loan) or an increase in clients' insolvency risk. As a matter of fact, both these risks are credit risk-related. For insurance companies, this can translate in higher payments on insured risks. This implies an exposure to liquidity risk and physical risk.

In addition to indirect risks, financial actors can also be impacted directly by sustainability-related risks. As an example, distressed commodity markets can result for both banks and insurance companies in a specific market risk due to the increase in the volatility of the value of the investment portfolios (when they are invested, at least in part, in commodities or in financial instruments having commodities as underlying assets). Similarly, possible unfair treatment of workers, discriminatory treatment of women or minorities (linked to social inequality as main sustainability-related factor) can rise a reputational risk and possibly the need of compensation due to proven responsibility (that is, in this latter case, a liability risk).

As it is shown in the taxonomy, sustainability-related risks typically result for banks and insurance companies in an increase in the risks already under management, such as credit risk, market risk, liquidity risk, liability risk, operational risk or reputational risk. This conclusion can have indeed significant consequences in terms of risk management practices for financial intermediaries. In fact, a strong argument can be made according to the idea that the correct management of the sustainability-related risks in the financial industry has to derive from a proper refinement of the existing frameworks, more than a complete change in paradigm.[23] In this respect, it seems nevertheless necessary to develop specific forward-looking approaches and methodologies able to cope with the lack of data and information on the specific relationship between sustainability-related and financial risks.[24]

The structure offered by the taxonomy in categorising sustainability-related risks and their impact on financial risks is likely a first-of-a-kind.

[23] Similar conclusions seem to emerge from the recent studies of the British Prudential Regulation Authority, PRA, and by the French Autorité de contrôle prudentiel et de resolution, ACPR (see PRA 2018 and ACPR 2019).

[24] For a wider discussion on this issue, see Chapter 4.

It suffers some limitations due to the lack of data on the significance and strength of the relations proposed and it is limited in scope, not including, inter alia, the role of households and of financial actors other than banks and insurance companies (e.g. investment funds). In this respect, we should be conscious of the fact that for a framework to be useful, it must have clear testable implications, so that the proposed patterns may be supported or refuted by data. Further empirical research will hence be needed to test the effectiveness on the ground. Nevertheless, the taxonomy can be considered a limited but concrete first step in better framing the *sustainability-financial risk nexus*.

1.5   Pricing the Sustainability-Related Risks and New Market Failures

As mentioned, the need for financial actors to systematically take into account sustainability-related risks in their core business is progressively becoming material. Nevertheless, this desirable new attention could also engender some negative side effects. When relevant, the full consideration of the sustainability-related risks by financial intermediaries in their risk management frameworks may have two possible outcomes: an adjustment in the pricing components of financial services (in particular as concerns credits and insurance services) and a reassessment of their risktaking strategies. The effects of these outcomes on the real economy will probably be uneven between geographies or economic sectors and, also depending on the progressive development and sophistication of the risk management practices, may be concentrated in the areas more affected (or expected to be more affected) by sustainability-related risks.

On the one hand, the pricing outcome may result in an increase in the cost of accessing financial services for economic agents in (some) proportion to their exposure to sustainability-related risks. This may be the case for example of companies operating in regions under increasing risk of hurricanes, which may need to face an escalation in the cost of insurance. Or for oil companies that may experience a substantial increase in the cost of accessing external financing due to limitations in availability of funds following policy decisions to discourage the use of fossil fuels. However, correctly pricing the incidence of sustainability-related risks on their financial risks is probably the most effective way for financial actors to be shielded from unexpected financial and economic losses. In addition, such a possible development would be in compliance with the principles and structures of existing prudential regulations and hence the one likely to be encouraged by policymakers in the years to come.

On the other hand, pricing adjustment may not be effective in the case of sustainability-related risks of significant magnitude. Following a reassessment of their risk-taking strategies, financial actors could eventually refuse to keep providing credit or insurance services to some of their existing and potential clients, in consideration of the high impact of sustainability-related risks on the financial risks they would need to bear. As a matter of fact, a number of sustainability-related risks may become uninsurable and a number of banks' clients may lose their creditworthiness due to sustainability-related factors. This can be the example of businesses located in areas increasingly hit by floods and hence subject to substantial degradation of their economic potential or households living in islands under the threat of the rise of sea level. New market failures can hence materialise in future as following a deeper assessment of the impact of sustainability-related risks on the different economic agents.

Even though it can be expected that pricing adjustments and market failures will be in many cases relatively small or even absent, this will still build a case for the need of periodically assessing the social impact of the management of sustainability-related risks by financial actors. In this respect, the problem could be exacerbated by the substantial lack of data and reliable information on the specific relationship linking sustainability-related and financial risks and the possible adoption, in particular in the short-term, of excessively precautionary approaches. Eventually, a specific policy intervention may also become necessary. This may be in the form of price control or cost support for the access to key financial services, promotion of ad hoc reinsurance schemes, more favourable fiscal treatment for stranded assets. As a matter of fact, these measures, which are limited to easing the possible impact of the side effects of the full consideration of sustainability-related risks on financial risks, can only supplement the wider policy strategies to foster a more sustainable society. In this respect, they may be considered by policymakers within the broad category of transition measures.

REFERENCES

Adams, Z., & Glück, T. (2015). Financialization in commodity markets: A passing trend or the new normal? *Journal of Banking & Finance, 60,* 93–111.

Alm, K., & Sievänen, R. (2013). Institutional investors, climate change and human rights. *Journal of Sustainable Finance and Investment, 3*(3), 177–183.

Autorité de contrôle prudentiel et de résolution (ACPR). (2019). *French banking groups facing climate change-related risks.* Paris.

Bank for International Settlements (BIS). (2020). *The green swan. Central banking and financial stability in the age of climate change.* Basel.

Bank of England (BoE). (2015, September). *Breaking the tragedy of the horizon— Climate change and stability.* Speech given by Mark Carney Governor of the Bank of England.

Berrou R., Ciampoli N., & Marini V. (2019a). Defining green finance: Existing standards and main challenges. In M. Migliorelli & P. Dessertine (Eds.), *The rise of green finance in Europe. Opportunities and challenges for issuers, investors and marketplaces.* Cham: Palgrave Macmillan.

Berrou, R., Dessertine, P., & Migliorelli M. (2019b). An overview of green finance. In M. Migliorelli & P. Dessertine (Eds.), *The rise of green finance in Europe. Opportunities and challenges for issuers, investors and marketplaces.* Cham: Palgrave Macmillan.

Brown, M. E., & Funk, C. C. (2008). Food security under climate change. *Science, 319*(5863), 580–581.

European Central Bank (ECB). (2019, May). *Climate change and financial stability.*

European Commission (EC). (2018a). *Action plan: Financing sustainable growth.* COM(2018) 97 final, Brussels.

European Commission (EC). (2018b). *A Clean Planet for all. A European strategic long-term vision for a prosperous, modern, competitive and climate neutral economy.* COM(2018) 773 final, Brussels.

European Commission (EC). (2019a). *The European Green Deal.* COM(2019) 640 final, Brussels.

European Commission (EC). (2019b). *Guidelines on non-financial reporting: Supplement on reporting climate-related information.* 2019/C 209/01, Brussels.

Friede, G., Busch, T., & Bassen, A. (2015). ESG and financial performance: Aggregated evidence from more than 2000 empirical studies. *Journal of Sustainable Finance and Investment, 5*(4), 210–233.

Hebb, T. (2013). Impact investing and responsible investing: What does it mean. *Journal of Sustainable Finance and Investment, 3*(2), 71–74.

Intergovernmental Panel on Climate Change (IPCC). (2018). *Special report. Global warning of 1.5 °C.*

International Finance Corporation (IFC). (2017). *Green finance. A bottom-up approach to track existing flows*. Washington, DC.

Lehner, O. M. (Ed.). (2016). *Routledge handbook of social and sustainable finance*. Abingdon: Routledge International Handbooks.

Migliorelli, M. (2018). Cooperative banking in Europe today: Conclusions. In M. Migliorelli (Ed.), *New cooperative banking in Europe. Strategies for adapting the business model post-crisis*. Cham: Palgrave Macmillan.

Migliorelli, M., & Dessertine, D. (Eds.). (2019a). *The rise of green finance in Europe. Opportunities and challenges for issuers, investors and marketplaces*. Cham: Palgrave Macmillan.

Migliorelli, M., & Dessertine, D. (2019b). From transaction-based to mainstream green finance. In M. Migliorelli & P. Dessertine (Eds.), *The rise of green finance in Europe. Opportunities and challenges for issuers, investors and marketplaces*. Cham: Palgrave Macmillan.

Network of Central Banks and Supervisors for Greening the Financial System (NGFS). (2019, April). *A call for action. Climate change as a source of financial risk*.

Prudential Regulation Authority (PRA). (2018, September). *Transition in thinking: The impact of climate change on the UK banking sector*.

Renneboog, L., ter Horst, J., & Zhang, C. (2008). Socially responsible investments: Institutional aspects, performance, and investor behaviour. *Journal of Banking & Finance, 32*(9), 1723–1742.

Tamazian, A., Chousa, J. P., & Vadlamannati, C. (2009). Does higher economic and financial development lead to environmental degradation: Evidence from the BRIC countries. *Energy Policy, 37*, 246–253.

Task Force on Climate-related Financial Disclosures (TCFD). (2017). *Recommendations of the Task Force on Climate-related Financial Disclosures*. Basel.

United Nations (UN). (2015). *Transforming our world: The 2030 agenda for sustainable development*.

United Nations Environment Programme (UNEP). (2016, September). *Definitions and concepts. Background note* (Inquiry Working Paper 16/13).

United Nations Environmental Programme Finance Initiative (UNEP FI). (2018, July). *Navigating a new climate: Assessing credit risk and opportunity in a changing climate*.

Vandekerckhove, W., Leys, J., Alm, K., Scholtens, B., Signori, S., & Schäfer, H. (Eds.). (2012). *Responsible investment in times of turmoil*. Dordrecht: Springer.

World Economic Forum (WEF). (2019). *The global risks report 2019*.

Ziolo, M., Filipiak, B. Z., Bąk, I., Cheba, K., Tîrca, D. M., & Novo-Corti, I. (2019). Finance, sustainability and negative externalities. An overview of the European context. *Sustainability, 11*(15), 4249.

Ziolo, M., & Sergi, B. S. (Eds.). (2019). *Financing sustainable development. Key challenges and prospects*. Cham: Palgrave Macmillan.

CHAPTER 2

The Impact of Climate Risks on the Insurance and Banking Industries

Giorgio Caselli and Catarina Figueira

Abstract It is now largely recognised that the global climate has changed since the pre-industrial period. While the role of financial institutions in the transition to a low-carbon economy has received increasing attention over time, more limited has been the evidence on how climate change might affect financial institutions' balance sheets. This chapter aims to redress this paucity of evidence by examining the impact of climate risks on the banking and insurance industries. To this purpose, it presents the main channels through which the physical, transition and liability risks of climate change might translate into financial risks for banks and insurance companies, along with the key data available to date. The extent

G. Caselli (✉)
Cambridge Judge Business School, Centre for Business Research,
University of Cambridge, Cambridge, UK
e-mail: g.caselli@cbr.cam.ac.uk

C. Figueira
Economic Policy and Performance Group, Cranfield School
of Management, Cranfield University, Cranfield, UK
e-mail: Catarina.figueira@cranfield.ac.uk

M. Migliorelli and P. Dessertine (eds.), *Sustainability and Financial Risks*, Palgrave Studies in Impact Finance, https://doi.org/10.1007/978-3-030-54530-7_2

to which climate risks might impair financial stability while causing new market failures is also discussed.

Keywords Banking · Climate change · Financial risk · Financial stability · Insurance · Market failure

2.1   Introduction

It is now largely acknowledged that the global climate has changed relative to the pre-industrial period. Nineteen of the twenty warmest years in history have all occurred since 2001, with 2016 ranking as the warmest year on record (NASA 2020). Although multiple lines of evidence exist that these changes have been affecting organisms and ecosystems, as well as human systems and well-being (IPCC 2018), more scarce has hitherto been the evidence on the impact that a changing climate might have on the financial system and its players. Until recently—at least before the Paris agreement was adopted in December 2015—the discussion around the link between climate change and financial institutions has tended to focus on the role they might play as catalysts for the transition to a low-carbon economy. Particular attention has been devoted to understanding how banks and other financial intermediaries might support a smooth and effective transition to a greener world in their function as providers of funds to the real economy. However, somewhat limited have been the efforts to uncover the financial risks that climate change might pose to financial institutions.

This dearth of evidence on the implications of climate risks for financial institutions may be problematic, as the COVID-19 pandemic that broke out at the beginning of 2020—as well as a number of other large-scale events—has vividly highlighted the negative effects that shocks external to the financial system might have for financial stability and the real economy. There is increasing recognition that severe weather events such as tornadoes, floods or droughts—whose frequency and magnitude have increased over the past years as a consequence of climate change—have the potential to translate into various financial risks for financial institutions, possibly undermining the overall stability and resilience of the financial system (Carney 2015).

Against this backdrop, this chapter aims to redress the paucity of dialogue about the implications of climate change for the financial sector by examining the effects of climate-related risks on insurance companies and banks. The chapter is organised as follows. Section 2.2 assesses the impact of climate risks on the insurance industry. It starts by providing an overall discussion about the various effects of climate change in this industry and then identifies the key risks and losses. It subsequently examines the challenges associated with insurers' business models and the pricing of climate-related risks. This section also discusses the importance of third-party liability risks and concludes with an overview of climate risk reinsurance. Section 2.3 addresses the effects of climate risks on the banking industry. The first part of the section presents the main data available to date on the financial risks that climate change is likely to create for banks and other financial intermediaries. The second part reviews the empirical literature on the pricing of climate risks by banks, while the final part explains how these risks might affect financial stability and contribute to new market failures. Section 2.4 discusses the problem of measuring the exposure of insurance companies and banks to climate risks and summarises the key data that are currently available. The last section concludes with some recommendations.

2.2 Impacts on the Insurance Industry

Overwhelmingly, it is now acknowledged that climate change is a reality that is having significant direct and indirect effects on society. It appears that these may become even more prevalent in the future. With climate change comes an increase in climate-related risks, mostly associated with the uncertainty surrounding the full impact of climate change and the difficulty in measuring such risks. This is indeed a very serious challenge faced by the insurance industry. As argued by Hecht (2008), *"if our society is to survive climate change without significant human costs, we must develop robust institutions and practices to manage these risks"*.

The insurance industry provides a significant service to companies, individuals, investors and other firms in the financial sector. By enabling the pooling of risk and savings, insurers spread policyholder risks, something that, individually, individuals and businesses would not be able to do. This is undoubtedly an important service which increases resilience across the wider economy, as for a fixed premium, often linked to long-term contracts, insurers provide certainty with respect to different financial outcomes, as noted by Swain and Swallow (2015).

One of the main purposes of the insurance industry is to match assets with liabilities, contributing to the financing of assets for infrastructure development and supporting the diversification across the financial system. It also provides income security to individuals through retirement products, income protection in case of unemployment and health care services. For instance, the industry paid over GBP5.7 billion in protection claims as per the Association of British Insurers (ABI 2019).

The abovementioned highlights the importance of the services provided by this industry across various sectors of the economy, providing an important contribution to economic growth. The role of insurers in supporting resilience across a range of economic activities is particularly important at a time of significant economic change. But it also stresses the impact that climate change can have on the efficiency of this industry and underlines some of its potential vulnerabilities, which, if not addressed properly, can impair the existence of this industry as we know it, with significant spillover effects to the wider society. Hence, it is important that an overview of the impact of climate change on the insurance industry considers both the implications to the industry with respect to the underwriting of climate change-related risks, investment activities, reporting and disclosure (CRO Forum 2019), as well as the economic and social role of the industry over the longer term (Bank of England 2015).

The above points also emphasise the significance of regulation in this matter. For example, this is the case of Solvency II—the European legislation Directive which came into force in January 2016. Summarily, Solvency II is a set of rules which should be adhered to by insurance companies and which focuses on how insurers should be funded and governed. It is based on three pillars (RIMES 2014):

- Pillar I covers requirements associated with the amount of capital an insurer should hold;
- Pillar II concentrates on governance, supervision and risk management requirements to ensure insurance firms are managed to, at least, a set standard;
- Pillar III addresses disclosure and transparency obligations, i.e. it sets out the necessary information that insurance firms need to disclose regarding their business.

This European Directive is particularly important in the context of ensuring the sustainability of financial institutions, namely insurers, and the strategic and operational decisions they take, as they face increased physical, transition and liability risks, as a result of the challenges climate change presents.

2.2.1   Types of Risks

Starting from the premise that insurance is, in essence, a mechanism for the transfer of risk in the context of the operation of markets, it is important to set out the main areas of risk for the industry: physical, transition and liability risks. In this subsection, we will focus on the first two areas of risk; we will address liability risks later in this chapter.

Physical risks relate to an increase in losses from climate trends or extreme weather events (Regelink et al. 2017). Climate trends include rise in average temperature, sea levels and coastal erosion, while floods and hurricanes are classified as extreme weather events, the accelerated frequency of which contribute to new, emerging physical risk trends. These risks are important, not only because they cause damage to property and often loss of lives, but also because they have an impact, for example, on the supply of resources, business operations and supply chains.

Transition risks relate to those risks which result from an attempt to reduce the transformational physical risks arising from climate change. According to the Bank of England (2019a), this type of risk requires a number of policy, market and technological changes to support possible financial costs and economic dislocations which may result from the process of reducing emissions and transitioning to a low-carbon economy. Among the sectors that are most exposed to this type of risk are those which relate to the extraction or production of fossil fuels and those that tend to emit large amounts of Greenhouse Gases (GHGs).

In addition, transition risks, together with changes in social behaviour, are expected to have a more general effect on many of the services and products that use fossil fuels, such as the sale of non-electric cars, properties that require a lot of energy due to limited insulation or even restaurants that do not offer vegan options. Infrastructure-related and utility businesses that also rely heavily on fossil fuels will most definitely be impacted by bans, carbon pricing and declining levels of demand in the future.

Importantly, there are positive implications to transition risks—as a result of a shift towards the use of increased levels of renewable energy, new opportunities for employment and growth in various sectors will emerge, as well as health benefits for the population. But, as argued by the European Academies' Science Advisory Council (EASAC 2018), there is "no silver bullet"—continuous research in new technologies requires large capital investments, can involve, in some cases, high investment risks and many of the new technologies also have significant limitations, such as high running costs. Adding to the impact of physical and transition risks are *third-party liability risks*. Their relevance for the insurance industry is paramount, but we will talk further about this in a separate subsection.

2.2.2   Uninsurable Risks and Losses

If we now concentrate on the impact that physical and transition risks can have on the insurance industry, we can clearly identify two areas that can shape the scenarios for which insurers need to prepare: one relates to the strength of response to the alleviation of climate change as per the Paris Agreement and the second concerns the pathway through which transition risks are being absorbed, i.e. how disruptive or smooth the transition to a low-carbon economy is established. The framework produced by the Network for Greening the Financial System (NGFS) and published by the UK Office for Budget Responsibility (2019) underscores the serious impact that these two types of risk can have for businesses and society at large, as can be seen in Fig. 2.1.

Specifically, the top right-hand corner box of this figure exacerbates the possible difficulty for insurers to insure certain risks properly and avoid significant losses. Indeed, it brings to the fore the potential for some risks to be regarded by the industry as non-insurable, in order to avoid very significant losses for the insurance business.

The latest developments regarding COVID-19 demonstrate how a virus has not only affected the health of so many individuals but has, in essence, shut down large parts of the world economy, with businesses closed, supply chains disrupted and a huge loss for both individuals and businesses. In such a case, the clauses associated with insurance contracts play an ever important role—for example, to what extent can an individual unemployment benefit insurance cover the damage caused by such a global, widespread health event?

PHYSICAL RISKS		
	Strength of response in meeting climate targets	
	met	not met
TRANSITION RISKS — Transition pathway — disorderly	*Disorderly* Sudden response is disruptive but able to meet climate targets	*Too little, too late* Sudden response is disruptive and not enough to meet climate targets
TRANSITION RISKS — Transition pathway — orderly	*Orderly* Measured reduction of emissions from now to meet climate targets	*Hot house world* Continued increase of emissions; not enough measures in place to minimise physical risks

Fig. 2.1 Illustration of the NGFS framework (*Source* Authors' elaboration based on the NGFS scenario analysis framework [Office for Budget Responsibility 2019])

The example of Flood Re, a publicly funded scheme in the UK, which was created to enable access to affordable insurance in areas prone to flooding also highlights the need for insurers to work closely with governments—it will continue to be the case (perhaps increasingly so) that some risks will be so expensive to incorporate in premiums that, unless governments are willing to bear some of the cost, insurance will not be a viable proposition in some areas and for some individuals and businesses.

More generally, physical and transition risks linked to climate change pose a number of very significant challenges to the sector. The Financial Stability Institute (FSI) of the Bank for International Settlements sets out a summary of these challenges (FSI 2019), which we present succinctly in five columns in Table 2.1 (related to insurance risk, market risk, credit risk, operational risk and liquidity risk) within the context of two examples.

2.2.3 Business Models and Pricing Adjustments

Given the fast developments that we are witnessing concerning the changes with the climate change and the resulting increased risks we have discussed earlier in this chapter, insurers and indeed the financial sector

Table 2.1 Challenges associated with physical and transition risks in the insurance industry

	Insurance risk	Market risk	Credit risk	Operational risk	Liquidity risk
Physical risk Example: Melting ice which increases sea levels and can cause floods	Higher than expected insurance claims payouts	Fall in equity values due to physical losses and business interruptions	Downgrade of credit rating of reinsurers, increasing the risk of losses	Physical damage to insurers' premises, disrupting the operation of the business	Higher policy cancellations to supplement lost income
Transition risk Example: Carbon tax, a government policy to reduce GHG emissions	Potential underpricing of new insurance products	Investment losses and lower asset values	Losses from corporate debt investments	Potential increased exposure to cyber risk	Pressure to invest in long-term green infrastructure projects

Source Authors' elaboration based on FSI (2019)

need to boost their climate readiness. This also means that their business models require reconsideration so that pricing of the products and services on offer can support insurers through the process of becoming more resilient within the context of emerging climate-related risks.

To be successful at identifying the key risks and opportunities for the sector and measure their effects appropriately, it is important that insurers work closely with policymakers towards an alleviation of climate risk exposure and collaborate with various other stakeholders to ensure that public policies which foster climate risk resilience are developed. In addition, insurers should also engage with rating agencies and experts in the field of environmental risk management to improve the accuracy of their pricing.

Importantly, as detailed by Deloitte (2019), the following steps will certainly prove useful in instilling climate risk readiness:

- Embedding the importance of climate risk within the business by, for example, linking executive compensation to performance metrics which should be closely linked to sustainability;
- Making use of advanced data analytics and engaging with the climate and data science research communities to improve the assessment of climate risk through developments in risk selection and pricing;
- Developing a holistic approach to climate risk exposure, by incorporating it in the insurers' enterprise risk management (ERM) framework. This will enable insurers to establish promptly correlations of certain impacts across both liabilities and investments.

In order to improve pricing strategies, firms will also be required to adjust their business models so that the input and output variables considered (as well as the way they are measured) incorporate the various scenarios as per the NGFS scenario analysis framework in Fig. 2.1. Specifically, with respect to the input variables, these should most definitely include macroeconomic variables, such as real GDP, inflation and unemployment rate and also financial variables which provide data on government bond yields, equity and commodity prices. However, central to the input variables is a set of climate variables that provides information on the expected frequency and impact of weather events, details carbon prices and provides a measure of emissions.

With respect to the output variables considered, firms need to work hard at sizing their risks and this certainly involves the development of an improved method in valuing assets and liabilities as a result of climate risks. With respect to general insurers' underwriting strategy, in particular, there are certain elements that are of utmost importance, such as knowledge about the location of the risk, the nature of the risk and the potential interconnection between risks—these are clearly central to their portfolio management.

Regarding life insurance, the specific combination of direct and indirect physical impacts as well as societal impacts needs to be addressed when dealing with these insurers' level of business exposure. Heatwaves, storms and floods are examples of direct physical impacts on life insurance but so are air pollution, environmental degradation, diseases which may have an indirect impact on the business. More widely, the state of public health infrastructure and political (in)stability in a country or region of the world represent societal impacts which are relevant to life insurance. In fact, all these factors ultimately may impact levels of mortality and morbidity and, hence, should be incorporated in the premium formulation, claims expectations and in the analysis of insurability of various strata across populations.

With respect to insurers' investment portfolios, two organisations, Investing Initiative (2II) and the ClimateWise Insurance Advisory Council have developed tools that focus on climate risk metrics. The latter, in particular, focuses on the quantification of transition risks for infrastructure investments, adopting the following steps: portfolio risk and opportunity exposure, asset impact identification and financial modelling analysis, as detailed in CRO Forum (2019).

In the meantime, insurers (general and life insurers) need to improve their modelling approach with respect to the assumptions used and advance data collection to improve forecasting. As per the World Bank (2016), climate change risks are often incorrectly priced due to mainly four factors: short-term approach to modelling of risks, inconsistent regulation across countries, asymmetric information, broad range of, for example, carbon prices across the world and lack of accurate data, which makes financial analysis very difficult and potentially less reliable.

Crucially, insurers need to recognise the challenges within their business models and ensure that management decisions and business models are aligned with the current expectations regarding the various climate scenarios (Bank of England 2019b).

Summarily, the insurance firms' ability to adjust their pricing to climate-related risks will certainly depend on a combination of operational, business and structural factors. Operational factors deal with what is known as the catastrophe risk modelling—complex models normally dealing with existing risks. Business model factors focus on the diversification of a range of risks, the transferring of some of the risk to reinsurers for risk mitigation purposes and the inverse production cycle. Finally, structural factors considered in the process of pricing should embed regulatory capital requirements and accommodate details relating to the duration of contracts.

2.2.4   Third-Party Liability Risks

It was mentioned earlier that third-party liability risks constitute another type of risk which needs to be considered by insurers when addressing the effects of climate change. These risks are central to the operation of insurers because they can have long-lasting implications for their businesses. Liability risks relate to the effect that can emerge sometime in the future if parties who have suffered loss seek compensation from those they believe are responsible for the damage inflicted on them. These risks need to be central to insurers' considerations when policies are formulated and sold because not only can they require a certain degree of speculation, but they can most certainly add to some disruption of the insurance business in the form of a potentially significant increase in claims over the longer term. This can present a serious financial problem for insurers, particularly if cover is proven for certain types of liability. For example, at the moment, insurers are dealing with serious liability challenges relating to the policies they have sold to businesses and individuals, as a result of COVID-19. Their response to events like these can lead to wider concerns for society and, in some cases, intensify the potential need to nationalise, to a certain extent, some physical and liability-related risks.

Third-party liability risks are often viewed as indirect risks as they relate to the impact that, for example, a flood can cause on certain business lines and the subsequent effect on third parties, who seek to recover losses from those they believe are responsible for these losses. For example, not only an extreme weather event may directly affect premises of a business and, if relevant, any products stored on-site, but it will contribute to financial loss and may have wider economic implications too—for example, it can affect one (or more) supply chain(s) and even cause displacement. Some of the most common examples of contracts that address third-party liabilities are professional indemnity or director's insurance contracts.

Generally, claims relating to this kind of risk in the context of climate change will focus on losses resulting from those who were insured and who failed to account for damage that they may have caused to the environment, known as "loss and damage" from climate change, where the impact of climate change has not been mitigated by taking the necessary steps to reduce, for instance, emissions, as per the United Nations Framework Convention on Climate Change (UNFCCC) Warsaw Agreement (UNFCCC 2014).

Third-party liability risks can also relate to a failure to comply with regulations. Claims due to asbestos use is one such example that has cost close to USD90 billion in claims in the United States alone. Another example is the long-term impact that pollution can have on various parties, such as individuals, past and current site owners of say potentially polluting businesses, impact on employees' health, etc.

With increased climate risks, it is expected that insurers will require to allocate what is already a large proportion of their balance sheet provisions to future and uncertain claims, on the basis of third-party liabilities. As of 2014, approximately 39% of total provisions in the insurance industry were related to addressing claims related to this type of risk—the largest percentage of provisions, followed by motor claims. Moving forward, insurers will be required to carefully consider all aspects of liability risks, in order to mitigate the physical impact of climate change they are liable to pay, including being able to address potential liabilities related to failure of businesses to adapt, examples of which can include governance issues or failure of clients to disclose or comply with the relevant legislation.

Finally, we will turn to another area of business in the insurance industry—reinsurance.

2.2.5   Reinsurance of Climate Risks

When we consider climate risks, we most certainly have to bear in mind the significant function played by reinsurance firms—their role in providing underwriting, pricing, claim management and general consultancy to primary insurance firms is of prime importance in absorbing shocks impacting the insurance industry (Upreti and Adams 2015). Among the concerns for reinsurers, climate risk has been recently ranked third in the Insurance Banana Skins ranking, as detailed in the latest report by PwC (2019), which provides evidence of its utmost importance to this area of business.

Climate change is indeed challenging the current reinsurance models due to two main factors: on the one hand, the risk of increased natural disasters will lead to significant claim charges and, on the other hand, it can impact the industry's reputational risk as they may, in fact, not be able to cover against, at least, some of emerging climate risks and hence reinsurers may no longer be in a position to offer certain solutions which could be instrumental to businesses in certain circumstances.

In essence, the above presents critical challenges to both sides of reinsurers' balance sheets (Swiss Re 2020). On the asset side, if reinsurers are to sustain their profitability, they need to carefully assess their investments in infrastructure funds and corporate bond holdings, as they may potentially become more exposed to certain physical and transition climate risks and, on the liability side, if they rely on existing models which are based on historical data, they may underestimate the premiums they have charged.

Therefore, in order to remain a sustainable industry, reinsurance firms need to invest in increasingly more sophisticated forward-looking business models, which consider all the relevant aspects to the socio-economic, technological, political and regulatory landscape and associated factors. Such models should also dynamically track the impact of climate change and, in particular, a warmer climate and, consequently, the derived additional exposures and vulnerabilities that result for the reinsurance industry.

2.3   Impacts on the Banking Industry

2.3.1   Lack of Data on the Impact of Climate Risks

The implications of climate risks for the banking industry have received increasing attention over the last decade, particularly since the adoption of the Paris Agreement in December 2015. Following the famous speech by Mark Carney—Governor of the Bank of England and former Chairman of the Financial Stability Board—on *"Breaking the Tragedy of the Horizon"* in September 2015 (Carney 2015), it is now largely accepted that climate change poses a number of financial risks for banks and other financial intermediaries. Many of these risks exhibit new characteristics, including greater scale, likelihood and interconnectedness (CCSF 2016). Collectively, these characteristics contribute to making climate risks more central in the current risk management agenda of banks and other financial institutions, as will be discussed later in this section and elsewhere in this book.

There tends to be agreement in the literature that risks associated with climate change do not represent a new category of risk for banks (Aubert et al. 2019). On the contrary, climate risks have the potential to manifest as types of risk already faced by banks and other financial intermediaries, namely credit, market and operational risks (Bank of England 2018). Table 2.2 provides an overview of the main ways in which climate risks, defined according to the taxonomy originally proposed by Carney (2015) and summarised in Sect. 2.2.1 above—i.e. physical, transition and liability risks—might translate into different types of financial risks for the banking industry.

Climate change might affect the quality of banks' credit portfolios through its effects on the ability of households and firms to repay their debts or meet their obligations. For example, physical risk arising from climate-related events such as droughts or long-term changes in precipitation could harm borrowers' income through decreased production capacity, translating into a higher probability of default and loss given default on loan books (Bolton et al. 2020). Credit risk might also increase as a result of the fall in collateral values and the write-off of assets located in regions at high climate risk. In turn, the transition towards a low-carbon economy could have a bearing on the riskiness of credit portfolios via changes in property values, which might stem from tighter energy efficiency standards or similar climate-related policy interventions (Monnin 2018). At the same time, the shift away from carbon-intensive sources of energy might imply that some assets could become stranded, impairing the value of banks' loan portfolios. This might be the case for corporate clients with business models that are not aligned with a 2 °C scenario (e.g. carbon extractors and emitters), whose earnings and business operations are likely to be the most affected by the transition—in particular if it happens lately and disorderly (Bank of England 2018). Although physical and transition risks can be regarded as the major sources of credit risk for banks and other financial intermediaries in relation to climate change, the quality of credit portfolios might also be weakened by greater liability risk. Insofar as compensation costs for climate-related losses or damages worsen borrowers' financial situation, liability risk on the part

Table 2.2 Main climate-related financial risks for the banking industry

	Credit risk	Market risk	Operational risk
Physical risk Direct losses caused by climate-related events • Acute (e.g. storms and floods) • Chronic (e.g. higher average temperatures and rising sea levels)	• Increase in default rates due to borrowers' declining revenue • Reduction in collateral values as a result of damages to property • Write-off of assets situated in high-risk areas	• Re-pricing of sovereign debt as a result of severe weather events	• Impact on business continuity (e.g. branches, infrastructure and staff) as a consequence of severe weather events
Transition risk Economic and financial consequences associated with the transition to a low-carbon economy	• Changes in property exposures as a result of stricter energy efficiency standards • Decline in value of loan portfolios due to stranded assets • Borrowers' losses arising from disruptive technology	• Re-pricing of securities and derivatives due to tighter climate-related policy	• Reputational risks associated with changing sentiment towards climate issues (e.g. divestment from fossil fuel companies)
Liability risk Liabilities arising from claims on climate-related losses or damages	• Increase in default rates due to borrowers' rising costs for climate change-related compensation		• Fines or penalties related to the consequences of climate change • Reputational risks stemming from a perceived inadequate response to climate change

Source Authors' elaboration based on Aubert et al. (2019) and Bank of England (2018)

of borrowers might transmit to banks' credit risk through an increase in default rates.

Together with their impact on credit risk, physical and transition risks have the potential to increase the market risk faced by banks and other financial intermediaries. Severe weather events such as hurricanes or floods could slow economic growth, for instance through large and sustained damage to national infrastructure. Greater levels of sovereign risk might lead to a re-pricing of national or local government's debt, possibly lowering the value of securities held on banks' balance sheets (Bank of England 2018). Market risk might also materialise as a consequence of stricter policies aimed at facilitating the transition to a low-carbon economy, which could cause a re-pricing of equities, corporate bonds and derivatives related to energy and commodities. Banks whose balance sheets are hit the hardest by credit and market risks could find it difficult to refinance themselves in the short-term, possibly facing liquidity risk and generating tensions in the interbank lending market (Bolton et al. 2020).

A third major category of bank risk that might be impacted by climate change is operational risk. Severe weather events such as storms or higher average temperatures are likely to disrupt business by causing damages to office premises and IT infrastructure or lowering staff's productivity and well-being. Additional impact on business continuity could be induced by higher volatility in the prices of inputs, including energy, water and insurance (Bank of England 2018). Another source of operational risk for banks and other financial intermediaries is tied to changes in policies and technologies as part of the adjustment towards a low-carbon economy. To the extent that this adjustment sparks a changing sentiment towards climate-related issues, such as increased pressure to divert capital away from fossil fuel companies and greater demand for green loans, climate change could represent a source of reputational risks for banks (TCFD 2017). These risks might be material if banks are perceived to be contributing to climate change or failing to manage climate-related risks, prompting claims by those who have suffered the losses or damages.

Although there is currently a good understanding of the channels whereby climate risks might affect banks and other financial intermediaries, a quantification of the likely impact of the physical, transition and liability risks of climate change on the banking industry is still underdeveloped (Summerhayes 2019). For instance, limited empirical data is available to date on the relationship between climate change and credit risk. Advancements in this area are hampered by a lack of historical data that banks can employ to evaluate the impact of climate risks on credit losses. For this reason, the quantification of physical, transition and liability risks tend to rely primarily on insights from climate scenarios and make the best use of expert judgements (Colas et al. 2018).

Theoretical support for the physical effects of climate change on the banking industry is provided by Dafermos et al. (2018), who establish that climate change is likely to increase the rate of default of corporate loans and threaten the stability of the banking system by destroying the capital of firms together with their profitability and liquidity. In addition, some preliminary evidence exists for the physical costs associated with climate-related events such as droughts and hurricanes having a negative impact on both equity and debt instruments through lower payoffs and higher non-performing loans (Campiglio et al. 2019).

In parallel to these studies, the last few years have witnessed a growing strand of research on the economic and financial consequences that the transition to a low-carbon economy might entail for the banking industry. Although this research is confronted with important data gaps and the need to rely on a number of assumptions, it offers some useful insights into the impact of transition costs on banks and other financial intermediaries.

One of the pioneering contributions is made by Battiston et al. (2017), who illustrate how their methodology can be used to perform a climate stress test of the banking system based on individual bank-level data. Their major conclusion, drawn from data for the top 50 listed European banks by total assets, is that banks would not default solely as a consequence of their loan exposures to firms in the fossil-fuel and utilities sectors. However, climate policies might cause significant volatility of large portions of banks' assets relative to their capital. A similar approach is taken by DNB (Vermeulen et al. 2018) for the Dutch banking sector, with their stress tests revealing that banks' losses are likely to reach 3% of total stressed assets in a disorderly energy transition. A large part of these losses is due to the interest rate effect associated with holdings of government bonds carrying longer maturities. It is also found that the regulatory capital (CET1) ratio of banks might fall by approximately 4% in a combined policy and technology shock scenario. These results are corroborated by a more recent study by Roncoroni et al. (2020), who develop a climate stress test framework to quantify the direct and indirect impact of a late and disorderly transition to a low-carbon economy. Focusing on the Mexican financial system as a laboratory, they find that an adverse scenario will generate systemic losses ranging between 2.5 and 4% of initial total assets—a sizeable amount.

2.3.2   Possible Pricing Adjustment Strategies

Insofar as climate change causes financial risks of the sort described in the previous subsection, it is important for banks and other financial intermediaries to account for these risks in their pricing strategies. As the Bank of Canada points out (Bank of Canada 2019, p. 29), "*[l]imited understanding and mispricing of climate-related risks could potentially increase the costs of transitioning to a low-carbon economy*". According to Thomä and Chenet (2017), the mispricing of climate risks has the potential to create a "carbon bubble"—i.e. an overvaluation of fossil fuel reserves and related assets that will materialise if the objective of containing climate change to well below 2 °C above pre-industrial levels is to be achieved (Schoenmaker and van Tilburg 2016). Moreover, if climate-related financial risks are being underestimated, capital is likely to be over-allocated to activities with higher risk. Alongside exposing creditors to potentially large losses, the underestimation of climate risks could result in central banks accepting collateral of insufficient credit quality (Monnin 2018). It follows that correctly pricing financial risks arising from climate change might support a more efficient allocation of capital by banks and other financial intermediaries, while ensuring they are not overexposed to risk (Chenet 2019).

Mispricing can occur for a variety of reasons, including limited data on carbon exposures, challenges of accounting for uncertain events in the future and discrepancy in time horizons—which means that households and firms that produce GHG emissions currently have no direct incentive to shift towards a low-carbon technology as they do not bear the damages or losses caused by their pollution (Thomä and Chenet 2017). For example, the lack of detailed and accurate information on climate risk at the level of individual assets and portfolios may hinder banks' ability to price risk and allocate capital properly (e.g. Monasterolo et al. 2017; Summerhayes 2019), implying that the efficient market hypothesis might not hold when it comes to climate change (Thomä and Chenet 2017). The consequences of incomplete information and ensuing mispricing of assets by banks and other market participants are summarised well by the Task Force on Climate-related Financial Disclosures (TCFD 2017, p. 1):

[I]nadequate information about risks can lead to a mispricing of assets and misallocation of capital and can potentially give rise to concerns about financial stability since markets can be vulnerable to abrupt corrections.

Campiglio et al. (2019) review the literature available to date and conclude that the impact of climate change on future financial asset performance will depend critically on the extent to which physical and transition costs are reflected in current asset prices. As Litterman (2011, p. 10) strongly emphasises in relation to pricing carbon emissions:

> Climate risk is not being priced. It should be priced immediately at a level that appropriately reflects fundamental uncertainty about catastrophic risks and a high level of societal risk aversion.

There is initial evidence in the literature to suggest that banks and other financial intermediaries have started to price in climate-related financial risks, yet not fully. Drawing on US data over the period 2001–2010, Cortés and Strahan (2017) show that small banks respond to local shocks created by exposure to natural disasters by increasing credit in affected areas and taking credit away from other areas. Small banks are found to mitigate the effects of credit reduction in connected markets by raising deposit rates in these markets to help finance additional lending. Further evidence from the US is provided by Jiang et al. (2019), who examine whether the risk associated with sea level rise has a bearing on the pricing of bank loans. They establish that the spreads for long-term loans—that is, loans with maturity longer than five years—go up with the sea level rise risk of the county where the borrower is located. In turn, Ouazad and Kahn (2019) identify a mispricing of assets vulnerable to natural disaster risk, i.e. guarantee fees associated with mortgage securitisation. They contend that the mispricing of mortgage risk carried in securitis- ers' balance sheets can represent a source of unhedged and unanticipated systemic risk.

Support for mispricing of climate-related financial risks in the banking industry also exists in relation to transition risk. Using the Clean Air Action launched by the Chinese Government in 2013 as a quasi- experiment, Huang et al. (2019) show that the loan spread charged to high-polluting firms increased by 5.5% after the policy implementation— compared with an increase of 50% in the default risk of these firms. This evidence suggests that banks may be pricing in climate-related transition risks, though not sufficiently. In a similar vein, Delis et al. (2019) focus on the syndicated loan market to investigate whether banks price in firms' polluting activities, i.e. stranded fossil fuel reserves. They find evidence consistent with banks charging significantly higher loan spreads to fossil fuel firms with greater exposure to climate policy risk, but only in the period after the Paris Agreement.

Further evidence that climate-related risks are not yet fully reflected in banks' prices is offered by a survey of 28 financial institutions in the Netherlands conducted by DNB (Regelink et al. 2017). The survey discovered that virtually none of these institutions were of the opinion that transition risks are adequately priced, indicating the potential for a sudden downwards shock in the banking industry arising from the introduction of new, low-carbon measures and technological developments.

Overall, the empirical evidence available so far concurs with the concern expressed by the NGFS that "*there is a strong risk that climate-related financial risks are not fully reflected in asset valuations*" (NGFS 2019, p. 4). For this reason, TCFD-style disclosure should be promoted further, as it is likely to help banks and other financial intermediaries adjust their pricing strategies in order to correctly account for climate-related risks.

2.3.3   Systemic Impact and Market Failures

The review of the literature presented in the previous subsection suggests that banks who fail to account for climate risks in the construction of their portfolios are pricing their holdings based on a misspecified model. To the extent that such mispricing has a bearing on the assets held by systemically important banks and other major financial intermediaries, there could be consequences in terms of systemic risk (Alessi et al. 2019). The link between mispricing of climate-related financial risks and the aggregate level of risk in the economy is corroborated by a recent analysis by the ECB (Giuzio et al. 2019). This analysis reveals that climate risks have the potential to become systemic for the euro area, particularly if banks and other financial institutions are not fully pricing in these risks.

According to a study by the ESRB (2016), an adverse scenario—whereby the transition to a low-carbon economy occurs late and abruptly—could have implications for systemic risk via three main channels: (i) lower energy supply and higher energy costs harm macroeconomic activity; (ii) banks and other financial institutions are negatively affected through their exposure to assets that are subject to a revaluation (e.g. carbon-intensive assets)[1]; (iii) the frequency and intensity of physical shocks (e.g. natural catastrophes) associated with climate change increase.

Climate risks may be regarded as systemic by nature, as they tend to impact the whole planet and are therefore non-diversifiable (Aglietta and Espagne 2016; DNB 2018). The unprecedented scale of these risks is such to have the potential to cause another major financial crisis (Saha and Viney 2019). In fact, one could argue that climate catastrophes are even more serious than most systemic financial crises, since they might pose an existential threat to humanity (Bolton et al. 2020).

Financial risks related to climate change have a number of distinctive elements, which are bound to give rise to considerable challenges for the banking industry. Among these elements is that they are far-reaching in breadth, that is, they will affect all economic agents (i.e. households, firms and governments) across all sectors and geographies (NGFS 2019). Therefore, their overall impact on the financial system is likely to be greater than other types of risks, while being potentially non-linear, correlated and irreversible (Bank of England 2018). Furthermore, despite uncertainty surrounding the precise outcome, there is a high degree of certainty that financial risks from climate change will occur sometime in the future (NGFS 2019). To use an expression introduced recently by the BIS (Bolton et al. 2020, p. 6), climate change can be viewed as "green swan" events:

[C]limate change represents a green swan: it is a new type of systemic risk that involves interacting, nonlinear, fundamentally unpredictable, environmental, social, economic and geopolitical dynamics, which are irreversibly transformed by the growing concentration of greenhouse gases in the atmosphere.

Empirical evidence on the implications of climate-related financial risks for bank soundness and financial stability already exists in the literature. Building on a sample covering 160 countries over the period 1997–2010, Klomp (2014) uncovers a positive relationship between natural disasters and the likelihood of a bank's default. Their analysis indicates that natural disasters may pose a substantial threat to the liquidity, yet not

---

[1] Although the carbon bubble alone is unlikely to be a source of systemic risk (Weyzig et al. 2014), it could combine with other sources of financial instability and create important destabilising effects for the financial system.

directly to the solvency, of the commercial banking sector. Similar results are obtained by Noth and Schüwer (2018), who investigate the effects of weather-related disasters on bank stability in the US between 1994 and 2012. They provide evidence that weather-related disasters indeed harm bank stability, as captured—among others—by significantly lower Z-scores, larger probabilities of default and higher non-performing asset ratios. The extent to which climate-related damages influence the stability of the banking system is explored further by Lamperti et al. (2019) using an agent-based climate-macroeconomic model. Their results show that climate change will lead to an increase in the frequency of banking crises (between 26 and 248%), with an additional fiscal burden of around 5–15% of GDP per year. It is estimated that approximately 20% of such effects will stem from the weakening of banks' balance sheets caused by climate change.

There is some preliminary evidence to suggest that bank stability might also be impacted by transition risk. Safarzyńska and van den Bergh (2017) establish that investments in renewable energy may lower interbank connectivity and lead to higher probability of bank failures. According to their analysis, financial stability may be hampered by a too quick transition to a low-carbon economy because the costs of financing investments in expensive renewable power plants may offset the enhanced profitability associated with existing gas power stations.

Besides its implications for bank soundness and financial stability, the literature available to date indicates that climate change might spawn new market failures. In fact, it could be maintained that climate change is itself the result of a market failure to account for the cost of GHG emissions to society (Fang 2018).

Drawing on US data during the 1990s, Garmaise and Moskowitz (2009) find that properties with greater exposure to hurricane risk are likely to receive less bank financing than other properties in the same zip code. This outcome translates into reduced provision of credit, limited participation of less wealthy investors and slower neighbourhood revitalisation in disadvantaged areas. Similar results for hurricane risk are presented by Brei et al. (2019), who show that deposit withdrawals together with a dry-up in non-deposit funding explain the contraction in bank lending that occurred in the two quarters following a hurricane in the Eastern Caribbean. At the same time, Duan and Li (2019) use mortgage origination as a laboratory to assess whether beliefs about climate change have a bearing on the decision-making of agents, establishing that abnormally high local temperatures reduce mortgage approval rates and loan amounts by 6.7%.

Additional evidence on the negative effects of weather-related disasters on bank lending is available for flood risk. Collier et al. (2013) collect data from a microfinance intermediary in Peru that is vulnerable to El Niño-related flood risk and conclude that loan losses caused by a natural disaster lead lenders to contract credit after the event—which in turn slows economic recovery for the affected area. These conclusions are confirmed by Choudhary and Anil (2017), who employ unprecedented flooding in Pakistan during 2010 as a natural experiment and show that banks disproportionately cut back on lending to new and less-educated borrowers following an exogenous shock to bank funding. They provide evidence consistent with this reduction in credit being driven by adverse selection and not being compensated by more lending by less-affected banks. Similarly, Faiella and Natoli (2019) investigate bank lending to non-financial firms at risk of flooding in Italy and find that banks tend to ration credit to firms with greater exposures to climate risk. Taken together, these results suggest that a rise in intermediation costs due to climate change-related events has the potential to create new market failures (or at least to amplify pre-existing ones).

2.4 THE PROBLEM OF MEASURING THE EXPOSURE TO CLIMATE-RELATED RISKS

There tends to be agreement in the literature that well-designed and effective climate-related disclosure is central to ensuring an orderly transition to a low-carbon economy (e.g. Batten et al. 2016). In June 2017, the TCFD published a set of recommendations concerning the voluntary disclosure of climate-related risks and opportunities by firms across all sectors, including insurance companies and banks. The rationale behind TCFD-style disclosure is encapsulated in the following statement accompanying the recommendations (TCFD 2017, p. ii):

> One of the essential functions of financial markets is to price risk to support informed, efficient capital-allocation decisions. Accurate and timely disclosure of current and past operating and financial results is fundamental to this function.

Although a growing number of financial institutions have made efforts to understand and implement the TCFD recommendations, a comprehensive assessment of their exposures to climate risks represents one of the most critical measurement gaps in relation to climate change (Giuzio et al. 2019). A survey by DNB (Regelink et al. 2017) revealed that transition risks are not yet fully incorporated into financial institutions' risk management frameworks, primarily because of scarce and incomplete information on their exposures to sectors with high levels of CO_2 emissions as well as on the energy labels of their real estate exposures. While sectoral analysis can offer a first approximation of financial institutions' exposures to climate risks, it abstracts from important differences in production processes and technologies within sectors (Giuzio et al. 2019). These problems might be compounded by the lack of granular data on the geographical location of their exposures. In addition, even if spatial data on their real estate exposures exists, it might not be systematically available in financial institutions' information systems (Aubert et al. 2019). All these factors contribute to making financial institutions' exposures to climate-related risks considerably hard to measure.

Nevertheless, the FSI has recently published a climate risk assessment with respect to the insurance industry (FSI 2019), based on a survey conducted by the Prudential Regulation Authority (PRA) to large insurers (including both life and non-life insurers) to establish to what extent they were exposed to two main risks: physical risks arising from climate change and other risks resulting from the transition to a low-carbon economy.

According to the PRA, the selected insurers were asked to comment on the impact of three scenarios on their models and also the effect of these on their asset valuations. The three scenarios considered were: A— a disorderly transition to a low-carbon economy as per the *IPCC Fifth Assessment Report* published in 2014; B—a long-term transition, with a maximum increase in temperature below 2 °C as set in the Paris Agreement and C—no transition, with a rise in temperature of approximately 4 °C above pre-industrial temperature levels by the year 2100. The results are presented in Tables 2.3 and 2.4.

Tables 2.3 and 2.4 provide some striking figures, particular with respect to Scenario C, where the impact on both insurers' liabilities and investments increases exponentially.

The even more limited evidence on climate risks available for the banking industry is somewhat concerning, given the significant size of financial institutions' exposures to these risks. Weyzig et al. (2014) find

Table 2.3   Impacts of physical risks on general insurers' liabilities

Sector	Assumptions	Physical risks scenario		
		A	B	C
US hurricane exposed lines of business	Percentage increase in frequency of major hurricanes	5%	20%	60%
	Uniform increase in wind speed of major hurricanes	3%	7%	15%
	Percentage increase in surface run-off resulting from increased tropical cyclone-induced precipitation	5%	10%	40%
	Increase in cm in average storm tide sea levels for US mainland coastline between Texas and North Carolina	10 cm	40 cm	80 cm
UK weather-exposed lines of business—flood, freeze and subsidence	Percentage increase in surface run-off resulting from increased precipitation	5%	10%	40%
	Uniform increase in cm in average storm tide sea levels for UK mainland coastline	2 cm	10 cm	50 cm
	Increase in frequency of subsidence-related property claims using as benchmark the worst year on record	3%	7%	15%
	Increase in frequency of freeze-related property claims using as benchmark the worst year on record	5%	20%	40%

Source Authors' elaboration based on FSI (2019)

that total equity, bond and credit exposures of EU banks to high-carbon assets corresponded to EUR460–480 billion (1.4% of their total assets) at the end of 2012, with corporate loans to fossil fuel companies contributing to almost two-thirds of this value. In a similar vein, DNB (Regelink et al. 2017) assesses Dutch financial institutions' exposures to transition-sensitive sectors (i.e. those that are responsible for the bulk of CO_2 emissions) and shows that 11% of their balance sheet was tied to carbon-intensive sectors in early 2017. This evidence is corroborated by the results of another study by DNB (Vermeulen et al. 2018), which conducts a climate stress test on over EUR2200 billion of assets held

Table 2.4 Impacts of risks on insurers' selected investments

Impacted sector	Investment portfolio in following sectors	Assumptions—change in equity value for sections of investment portfolio comprising material exposure to the energy sector as per below	Transition risks scenario			Physical risks scenario		
			A	B	C	A	B	C
Fuel extraction	Gas/coal/oil (including crude)	Coal	−45%	−40%				
		Oil	−42%	−38%				
		Gas	−25%	−15%				
							−5%	−20%
Power generation	Power transmission and delivery of natural gas and renewables (production and transmission)	Coal	−65%	−55%				
		Oil	−35%	−30%				
		Gas	−20%	−15%				
		Renewables (including nuclear)	+10%	+20%				
							−5%	−20%

Source Authors' elaboration based on FSI (2019)

by Dutch banks, insurers and pension funds and concludes that banks are the most exposed to carbon-intensive industries—with a total exposure of 13% against a figure of 5% for insurers and 8% for pension funds. Moreover, the ECB (Giuzio et al. 2019) establishes that Euro area banks' exposures to firms contributing to carbon emissions are sizeable, with the 20 largest emitters accounting for approximately 20% of total large exposures (1.8% of total assets of the sample banks). Data from the Banque de France (Aubert et al. 2019) also indicate that total exposure of the largest banking institutions in France to sectors that are most important GHG emitters reached 12.2% of total credit risk exposures at the end of December 2017, while banks' exposures to climate policy-relevant sectors represent a portion of loan portfolios comparable to their capital (Battiston et al. 2017)—raising serious concerns if a substantial part of these portfolios ends up as stranded assets.

2.5 CONCLUSIONS

The evidence reviewed in this chapter suggests that climate change might represent a source of significant financial risks for insurance companies and banks and that these risks are likely to increase in the years ahead. Since the literature indicates that risks associated with climate change have the potential to become systemic, it is critical for insurance companies and banks to identify, price and manage these risks appropriately.

There is now growing awareness on the part of financial institutions about the importance of correctly quantifying climate-related risks. It is encouraging to see the results of surveys by central banks and other financial authorities showing that an increasing number of insurance companies and banks are addressing climate risks at the group strategy level, rather than simply as a concern of their CSR function (e.g. Aubert et al. 2019; Bank of England 2018). Survey results also reveal that insurance companies and banks have started to assess the opportunities that might be brought about by the transition to a low-carbon economy, including the development of new products and the support to customers throughout the transition period (APRA 2019).

However, the long-term horizon that distinguishes climate-related risks makes it extremely challenging for insurance companies and banks to identify, measure and monitor these risks. These challenges are compounded by considerable data gaps, particularly in relation to financial institutions' exposures to the physical, transition and liability risks created by climate change. For example, there is evidence that transition risks are not yet fully incorporated into financial institutions' risk management frameworks, largely because of limited information on key aspects of their portfolio exposures such as the levels of CO_2 emissions and the energy labels of real estate properties (Regelink et al. 2017).

It follows that more efforts need to be made to mitigate the potential impact of climate risks on financial stability, while ensuring that financial institutions' practices are aligned with the target of limiting global temperature rises to well below 2 °C. The complex nature of climate change requires coordinated actions by a multitude of players, including financial institutions, the private sector and financial authorities. It is recommended that insurance companies and banks integrate financial risks from climate change into their risk management frameworks and discuss them at the board level. Climate-related financial risks should be identified and addressed at the earliest possible stages in order to reduce their effects on financial institutions' balance sheets. Insurance companies and banks should be supported in this endeavour through better availability and comparability of data on their exposure to climate-related risks. To this end, disclosure in line with the TCFD recommendations should be encouraged further. In turn, financial authorities such as central banks and other supervisors have a major role to play in ensuring that climate risks are effectively taken into account by insurance companies and banks. Since climate change presents a number of financial risks that are relevant to supervisory authorities, greater emphasis might be placed on strengthening current approaches to supervision through the integration of climate risks into prudential regulation requirements. The debate on whether financial institutions should be required to hold additional capital in view of their climate risks is one that is likely to attract continued interest in the time to come.

References

Aglietta, M., & Espagne, É. (2016). *Climate and finance systemic risks, more than an analogy? The climate fragility hypothesis* (CEPII Working Papers, 10).

Alessi, L., Ossola, E., & Panzica, R. (2019). *The Greenium matters: Evidence on the pricing of climate risk*. Luxembourg: Publications Office of the European Union.

Association of British Insurers (ABI). (2019). *Record 98.3% of protection claims paid out in 2019*. Available at https://www.abi.org.uk/news/news-articles/2020/05/record-98.3-of-protection-claims-paid-out-in-2019/. Accessed 23 May 2020.

Aubert, M., Bach, W., Diot, S., & Vernet, L. (2019). *French banking groups facing climate change-related risks* (Banque de France Analyses et Synthèses, 101).

Australian Prudential Regulation Authority (APRA). (2019). *Climate change: Awareness to action*. Available at https://www.apra.gov.au/sites/default/files/climate_change_awareness_to_action_march_2019.pdf. Accessed 7 April 2020.

Bank of Canada. (2019). *Financial system review—2019*. Available at https://www.bankofcanada.ca/wp-content/uploads/2019/05/Financial-System-Review—2019-Bank-of-Canada.pdf. Accessed 31 March 2020.

Bank of England. (2015, September). The *impact of climate change on the UK insurance sector*. A climate change adaptation report by the Prudential Regulation Authority.

Bank of England. (2018). *Transition in thinking: The impact of climate change on the UK banking sector*. Available at https://www.bankofengland.co.uk/-/

media/boe/files/prudential-regulation/report/transition-in-thinking-the-impact-of-climate-change-on-the-uk-banking-sector.pdf. Accessed 26 March 2020.

Bank of England. (2019a, March 21). *A new horizon.*

Bank of England. (2019b, December). *The 2021 biennial exploratory scenario on the financial risks from climate change* (Discussion Paper).

Batten, S., Sowerbutts, R., & Tanaka, M. (2016). *Let's talk about the weather: The impact of climate change on central banks* (Staff Working Papers, 603).

Battiston, S., Mandel, A., Monasterolo, I., Schütze, F., & Visentin, G. (2017). A climate stress-test of the financial system. *Nature Climate Change, 7*(4), 283–290.

Bolton, P., Despres, M., Pereira Da Silva, L. A., Samama, F., & Svartzman, R. (2020). *The green swan: Central banking and financial stability in the age of climate change.* Available at https://www.bis.org/publ/othp31.pdf. Accessed 31 March 2020.

Brei, M., Mohan, P., & Strobl, E. (2019). The impact of natural disasters on the banking sector: Evidence from hurricane strikes in the Caribbean. *The Quarterly Review of Economics and Finance, 72*, 232–239.

Cambridge Centre for Sustainable Finance (CCSF). (2016). *Environmental risk analysis by financial institutions: A review of global practice.* Cambridge, UK: Cambridge Institute for Sustainability Leadership.

Campiglio, E., Monnin, P., & von Jagow, A. (2019). *Climate risks in financial assets* (CEP Discussion Notes, 2).

Carney, M. (2015). *Breaking the tragedy of the horizon—Climate change and financial stability.* Available at https://www.bankofengland.co.uk/-/media/boe/files/speech/2015/breaking-the-tragedy-of-the-horizon-climate-change-and-financial-stability.pdf. Accessed 26 March 2020.

Chenet, H. (2019). *Climate change and financial risk* (Unpublished manuscript).

Choudhary, M., & Anil, J. (2017). *Finance and inequality: The distributional impacts of bank credit rationing* (International Finance Discussion Papers, 1211).

Colas, J., Khaykin, I., Pyanet, A., & Westheim, J. (2018). *Extending our horizons: Assessing credit risk and opportunity in a changing climate: Outputs of a working group of 16 banks piloting the TCFD recommendations.* Available at https://www.unepfi.org/publications/banking-publications/extending-our-horizons/. Accessed 31 March 2020.

Collier, B. L., Miranda, M. J., & Skees, J. R. (2013). *Natural disasters and credit supply shocks in developing and emerging economies* (Working Papers, 3). Wharton Risk Management and Decision Processes Center.

Cortés, K. R., & Strahan, P. E. (2017). Tracing out capital flows: How financially integrated banks respond to natural disasters. *Journal of Financial Economics, 125*(1), 182–199.

CRO Forum. (2019). *The heat is on. Insurability and resilience in a changing climate* (Emerging Risk Initiative—Position Paper).

Dafermos, Y., Nikolaidi, M., & Galanis, G. (2018). A stock-flow-fund ecological macroeconomic model. *Ecological Economics, 131,* 191–207.

Delis, M., de Greiff, K., & Ongena, S. (2019). *Being stranded with fossil fuel reserves? Climate policy risk and the pricing of bank loans* (Swiss Finance Institute Research Paper Series, 10).

Deloitte. (2019). *Climate risk: Regulators sharpen their focus. Helping insurers navigate the climate risk landscape.* Deloitte Center for Financial Services.

De Nederlandsche Bank (DNB). (2018). *DNB sustainable finance platform's working group on climate risk.* Available at https://www.dnb.nl/en/binaries/rapport_tcm46-374715_tcm47-379199.pdf. Accessed 31 March 2020.

Duan, T., & Li, F. W. (2019). *Climate change concerns and mortgage lending* (Unpublished manuscript).

European Academies' Science Advisory Council (EASAC). (2018). *Negative emission technologies: What role in meeting Paris Agreement targets?* (EASAC Policy Reports, 35).

European Systemic Risk Board (ESRB). (2016). *Too late, too sudden: Transition to a low-carbon economy and systemic risk* (Reports of the Advisory Scientific Committee, 6).

Faiella, I., & Natoli, F. (2019). *Climate change and bank lending: The case of flood risk in Italy* (Unpublished manuscript).

Fang, C. C. (2018). Carbon pricing: Correcting climate change's market failure. *Sustainability: The Journal of Record, 11*(4), 162–166.

Financial Stability Institute (FSI). (2019). *Turning up the heat—Climate risk assessment in the insurance sector* (FSI Insights on Policy Implementation, 20).

Garmaise, M. J., & Moskowitz, T. J. (2009). Catastrophic risk and credit markets. *The Journal of Finance, 64*(2), 657–707.

Giuzio, M., Krusec, D., Levels, A., Melo, A. S., Mikkonen, K., & Radulova, P. (2019, May). Climate change and financial stability. In European Central Bank (ECB) (Eds.), *Financial Stability Review* (pp. 120–133).

Hecht, S. B. (2008). Climate change and the transformation of risk: Insurance matters. *UCLA Law Review, 55,* 1559–1620.

Huang, B., Punzi, M. T., & Wu, Y. (2019). *Do banks price environmental risk? Evidence from a quasi natural experiment in the People's Republic of China* (ADBI Working Paper Series, 974).

Intergovernmental Panel on Climate Change (IPCC). (2018). *Global warming of 1.5 °C: An IPCC special report on the impacts of global warming of 1.5 °C above pre-industrial levels and related global greenhouse gas emission pathways, in the context of strengthening the global response to the threat of climate change, sustainable development, and efforts to eradicate poverty.* Available at https:// www.ipcc.ch/sr15/download/#full. Accessed 9 May 2020.

Jiang, F., Li, C. W., & Qian, Y. (2019). *Can firms run away from climate-change risk? Evidence from the pricing of bank loans* (Unpublished manuscript).

Klomp, J. (2014). Financial fragility and natural disasters: An empirical analysis. *Journal of Financial Stability, 13,* 180–192.

Lamperti, F., Bosetti, V., Roventini, A., & Tavoni, M. (2019). The public costs of climate-induced financial instability. *Nature Climate Change, 9*(11), 829–835.

Litterman, R. (2011). Pricing climate change risk appropriately. *Financial Analysts Journal, 67*(5), 4–10.

Monasterolo, I., Battiston, S., Janetos, A. C., & Zheng, Z. (2017). Vulnerable yet relevant: The two dimensions of climate-related financial disclosure. *Climatic Change, 145*(3–4), 495–507.

Monnin, P. (2018). *Integrating climate risks into credit risk assessment: Current methodologies and the case of central banks corporate bond purchases* (CEP Discussion Notes, 4).

NASA. (2020). *Global temperature.* Available at https://climate.nasa.gov/vital-signs/global-temperature/. Accessed 8 May 2020.

Network for Greening the Financial System (NGFS). (2019). *A call for action: Climate change as a source of financial risk.* Available at https://www.ngfs.net/sites/default/files/medias/documents/synthese_ngfs-2019_-_170 42019_0.pdf. Accessed 26 March 2020.

Noth, F., & Schüwer, U. (2018). *Natural disaster and bank stability: Evidence from the U.S. financial system* (SAFE Working Paper Series, 167).

Office for Budget Responsibility. (2019, July). *Fiscal risks report* (CP131).

Ouazad, A., & Kahn, M. E. (2019). *Mortgage finance in the face of rising climate risk* (NBER Working Papers, 26322).

PwC. (2019). *Reinsurance: Banana Skins 2019.* Centre for the Study of Financial Innovation (CSFI).

Regelink, M., Reinders, H. J., Vleeschhouwer, M., & van de Wiel, I. (2017). *Waterproof? An exploration of climate-related risks for the Dutch financial sector.* Available at https://www.dnb.nl/en/binaries/Waterproof_tcm47-363 851.pdf. Accessed 26 March 2020.

RIMES. (2014). *Solvency II: The data challenge* (White Paper).

Roncoroni, A., Battiston, S., Farfán, L. O. L. E., & Jaramillo, S. M. (2020). *Climate risk and financial stability in the network of banks and investment funds* (Unpublished manuscript).

Safarzyńska, K., & van den Bergh, J. C. J. M. (2017). Financial stability at risk due to investing rapidly in renewable energy. *Energy Policy, 108,* 12–20.

Saha, S., & Viney, B. (2019). How climate change could spark the next financial crisis. *Journal of International Affairs, 73*(1), 205–216.

Schoenmaker, D., & van Tilburg, R. (2016). What role for financial supervisors in addressing environmental risks? *Comparative Economic Studies, 58*(3), 317–334.

Summerhayes, G. (2019). Financial exposure: Climate data deficit. *Banque de France Financial Stability Review, 23,* 49–56.

Swain, R., & Swallow, D. (2015). The prudential regulation of insurers under Solvency II. *Bank of England Quarterly Bulletin, 55*(2). Available at https://www.bankofengland.co.uk/quarterly-bulletin/2015/q2/the-prudential-regulation-of-insures-under-solvency-2. Accessed 23 May 2020.

Swiss Re. (2020). *Sigma. Natural catastrophes in times of economic accumulation and climate change* (Swiss Re Institute Report 02/20).

Task Force on Climate-related Financial Disclosures (TCFD). (2017). *Final report: Recommendations of the task force on Climate-related financial disclosures.* Available at https://www.fsb-tcfd.org/wp-content/uploads/2017/06/FINAL-TCFD-Report-062817.pdf. Accessed 6 April 2020.

Thomä, J., & Chenet, H. (2017). Transition risks and market failure: A theoretical discourse on why financial models and economic agents may misprice risk related to the transition to a low-carbon economy. *Journal of Sustainable Finance and Investment, 7*(1), 82–98.

United Nations Framework Convention on Climate Change (UNFCCC). (2014). *Report of the conference of the parties on its nineteenth session.* Framework Convention on Climate Change.

Upreti, V., & Adams, M. (2015). The strategic role of reinsurance in the United Kingdom's non-life insurance market. *Journal of Banking & Finance, 61,* 206–219.

Vermeulen, R., Schets, E., Lohuis, M., Kölbl, B., Jansen, D.-J., & Heeringa, W. (2018). *An energy transition risk stress test for the financial system of The Netherlands* (DNB Occasional Studies, 7).

Weyzig, F., Kuepper, B., van Gelder, J. W., & van Tilburg, R. (2014). *The price of doing too little too late: The impact of the carbon bubble on the EU financial system* (Green New Deal Series, 11).

World Bank. (2016). *State and trends of carbon pricing.* Washington, DC: World Bank.

CHAPTER 3

Stranded Assets and the Transition to Low-Carbon Economy

Olaf Weber, Truzaar Dordi, and Adeboye Oyegunle

Abstract In the context of the low-carbon transition, stranded assets can be defined as assets that have suffered unanticipated or premature write-downs, devaluations or conversions to liabilities. These assets may refer to resource reserves, infrastructure or industries that may be affected by economic, physical or political changes along a pathway of decarbonisation. This chapter first gives a historical account of stranded assets in a low-carbon transition and presents a systematic review of the literature. Then, it proposes a comprehensive approach to understanding the multitude of factors resulting in stranded asset risk, by including case studies to show how responses to stranded asset risks vary by region. Finally, it offers

O. Weber (✉) · T. Dordi · A. Oyegunle
University of Waterloo, Waterloo, ON, Canada
e-mail: oweber@uwaterloo.ca

T. Dordi
e-mail: truzaar.dordi@uwaterloo.ca

A. Oyegunle
e-mail: adeboye.oyegunle@uwaterloo.ca

© The Author(s) 2020   63
M. Migliorelli and P. Dessertine (eds.), *Sustainability and Financial Risks*, Palgrave Studies in Impact Finance,
https://doi.org/10.1007/978-3-030-54530-7_3

a research agenda for future studies, addressing some of the limitations to current research.

Keywords Stranded assets · Low-carbon transition · Climate change · Transition risk · Sustainability-elated risks · Climate-related risks

3.1   Introduction

In the context of the low-carbon transition, stranded assets are defined as assets that have suffered unanticipated or premature write-downs, devaluations or conversions to liabilities (Ansar et al. 2013). These assets may refer to resource reserves, infrastructure or industries that may be affected by economic, physical or political changes along a pathway of decarbonisation.

The conditions that result in asset stranding have affected incumbent industries for centuries, from the early industrial revolution to advancements in computing. Large disruptive innovations such as the advent of rail, mainstreaming of telecommunications and personal computing have brought economic development and value creation but also the obsolescence of incumbent players. Modern examples from the digitisation of photography to video streaming have left monolithic corporations like Kodak and Blockbuster behind. The Schumpeterian notion of creative destruction is rich with such examples. However, stranded assets associated with the low-carbon transition are also driven by a social and political urgency to transition to a low-carbon economy.

However, "Stranded assets" is a relatively nascent term in the energy sector. The majority of publications on the topic are from after 2015. Notably, the financial risk is still under-represented in the literature. The financial risk associated with stranded assets is difficult to identify and measure because numerous conditions affect the outcome, and equally, multiple asset types can be affected. Stranded assets may arise from physical or transitional factors, ranging from extreme weather events, climate-related government policy and greater social awareness. Respectively, the financial risk varies by asset types at a regional and industry level. Responses to stranded assets may differ in developing and developed countries, and between oil-rich nations and island states. Beyond fossil fuel production, industries like agriculture and real estate may also be affected by extreme weather events like droughts or floods. Due to the numerous scenarios by which stranded assets may manifest in the economy, stranded asset research may benefit from a clear typology on how to identify and measure stranded asset risk.

This chapter is structured as follows. First, Sect. 3.2 gives a historical account of stranded assets in a low-carbon transition. We examine the early pressures that led to a discourse on stranded asset risk and the role of "unburnable carbon" as a catalyst for the institutionalisation of the topic. Section 3.3 then presents a systematic review of the current state of the literature on stranded assets. This is followed by Sect. 3.4, proposing a more comprehensive approach to understanding the multitude of factors resulting in stranded asset risk. Section 3.5 presents some case studies on how responses to stranded asset risks vary by region, comparing them to the proposed typology to identify limits to each model. Finally, Sect. 3.6 offers a research agenda for future study, addressing some of the limitations to current research.

3.2 The History of Stranded Assets

Early reference of stranded assets in academic literature dates back to the 1990s in the context of the devaluation of productive assets through the restructuring of electric utilities in the United States and Europe (Meyer 1997; Rothkopf 1997; Woychik 1995). The deregulation of electric utilities would allow for greater competition, which has the potential to cut prices for consumers; however, the value of incumbent utilities may also decline as contracts that were once economically viable are undercut by private electricity providers. Electric utilities thus refer to the difference in the value of contracts following restructuring as 'stranded costs' and literature on the topic asks whether utilities should be compensated for stranded costs and at what valuation.

The origins of stranded assets are rooted in the Schumpeterian (1942) notion of creative destruction, driven by a socio-technical energy transition away from fossil fuel production (Knuth 2017). There are examples of this phenomenon, as advancements in personal computing and digitisation revolutionised industries in recent decades (Green and Newman 2017). Much like the industrial revolutions of the past and advent of rail, electricity, oil and telecommunications brought the obsolescence of certain companies, infrastructure and capital. It also brought innovation, economic growth and value creation (Perez 1985, 2009). These innovations began as a niche but superior offerings to incumbent services, grow exponentially in their uptake and consequently create stranded assets (Green and Newman 2017).

Carbon-intensive fossil fuel production is now at risk of the same stranded assets that have reshaped industries before. As increasingly accessible low-carbon technologies such as solar and battery storage prove to be both superior in performance, and at a lower price point, incumbent producers are at risk of becoming obsolete. The advent of stranded assets through a low-carbon transition, however, differs from other examples—as the low-carbon paradigm is also driven by an environmental crisis. Thus, a myriad of factors, including technological change, government policy, greater social awareness and increasing climate-related extreme weather events, collectively influence the demand of fossil fuels (Green and Newman 2017). Under the trope of "leaving fossil fuels in the ground", these factors represent an interrelated devaluation effort against fossil fuels (Knuth 2017).

3.2.1   *The Low-Carbon Transition and "Unburnable" Carbon*

Stranded assets associated with a low-carbon transition are unique in that they are not strictly driven by technological innovation, but rather a need to limit carbon emissions to mitigate the worst effects of climate change. While the discourse around emissions reductions spans decades (Randalls 2010), the discourse involving the "low-carbon transition" and "stranded assets" can be directly traced to relatively recent work by Meinshausen et al. (2009) and Allen et al. (2009), on global carbon budgets. These works changed the narrative on emissions from that of incremental reduction to a ceiling limit on emissions; and in turn, laid the foundation for a "carbon constrained" future driven by a socially imposed limit to carbon production (Jaccard et al. 2018). This may be framed as a zero-sum choice, to either decarbonise by keeping fossil fuel resources in the ground and risking trillions in stranded assets, or to face a climate catastrophe. The emphasis on limits has significant implications—emissions must be eliminated within decades (Rogelj et al. 2018), and as such, the majority of economically proven fossil fuel reserves cannot be developed (Meinshausen et al. 2009). These unusable reserves could in-turn, be "stranded", losing value due to unanticipated or premature write-downs, devaluations or conversions to liabilities (Ansar et al. 2013).

The two-degree target that was adopted at the Copenhagen COP Summit in 2009 further cemented the urgency of a low-carbon transition. The target shifted the narrative from incremental emission reductions to a global limit on cumulative emissions, which, if surpassed, would likely exceed the globally accepted two-degree target. The target also shifted attention away from demand-side efficiencies (i.e. more fuel-efficient combustion vehicles) to supply-side constraints targeted squarely at the fossil fuel industry. Given that emissions from fossil fuel combustion are the dominant source of anthropogenic emissions (Quéré et al. 2013) and that these emissions are highly concentrated among some of the largest fossil fuel companies (Heede 2014), the low-carbon paradigm asserts that to limit carbon emissions, fossil fuel companies must reduce and eventually cease all production of their carbon reserves (Arbuthnott and Dolter 2013). This puts the brunt of stranded asset risk on carbon-intensive fossil fuel production.

The stranded assets discourse was institutionalised by the Paris Agreement. Around this time, investment banks issued warnings to investors on stranded carbon assets and (to a lesser degree) the carbon bubble, as did the world's largest hedge fund, Blackrock. It was at this time that the Bank of England's Governor Mark Carney warned that the carbon bubble would pose a systemic risk to the financial system, justifying regulation under the Bank of England's mandate (Carney 2015). This not only aligns with the rapid uptake of academic publications on the topic of stranded assets post-2015, but also with notable shifts in capital markets, as once ambivalent financiers have begun distancing themselves from financing, investing in and underwriting the fossil fuel industry (Strauch et al. 2020).

3.3 THE STATE OF LITERATURE ON STRANDED ASSETS

A systematic review of the literature indicates how the study has involved, latent topics of interest, and current limits to under understanding of stranded asset risk. We conduct our search of relevant literature on the *Web of Science* and *Scopus* database, using queries related to "stranded asset*", "asset strand*" and "stranded fossil fuel asset*" to identify pertinent publications. Our initial query indexed 102 publications on *Web of Science*, and 161 publications on *Scopus*.

The study of stranded assets in relation to the low-carbon transition is nascent. 90% of the publications in our sample were published after

2015. There has, however, been a notable evolution in the literature over time. Publications before 2009 referred almost exclusively to the restructuring of electric utilities. Only early work by Odenberger and Johnsson (2007) had begun examining stranded assets in the context of emissions reduction. In particular, they examined how low-carbon investment in UK electric generation could contribute to emission reduction targets and how continued investments in carbon-based long-term capital stock could result in costly early retirement of power plants to comply with (Kyoto's) emission goals. The discourse around carbon capture and sequestration and gained prominence in 2009, as a potential solution to the risk of stranded assets (Odenberger et al. 2009; Pearson et al. 2009). The disparity in emissions reduction between developed and developing countries also grew apparent at this time. A scenario analysis by Richels et al. (2009) considered how future emission reductions targets affected technological investment decisions in developed and developing countries, as a means to mitigate stranded assets. Stern (2012) considered stranded assets in the context of water, whereby high-cost reservoirs and desalination facilities could face asset stranding if competition from imported water is introduced.

Since 2015 we see the most recent evolution in stranded asset discourse (Fig. 3.1)—one that revolves primarily around global climate change and specifically, carbon-intensive fossil fuel production. An inductive analysis of prominent keywords alludes to some relevant topics. Out of the articles in the sample that were published after 2015, "climate change" and "fossil fuels" are referred to in 39 publications each. The discourse around risk and policy is also prevalent in the sample, with 43 publications referring to risk and 64 publications referring to policy. Finally, 20 publications refer to scenarios, and nine publications refer to divestment.

3.3.1   Factors of Stranded Assets Risk

In line with the purpose of this book, we delve deeper into how stranded asset "risk" is developed in the literature. Risk of stranded assets is applied in the context of carbon risk, financial risk, political risk and systematic risk. We examine select publications for illustrative purposes.

The most common application of stranded asset risk in the literature is in the context of risk associated with carbon and climate change. The global carbon budget is the driving factor of stranded asset risk and is thus extensively analysed across numerous contexts. Bang and Lahn (2019),

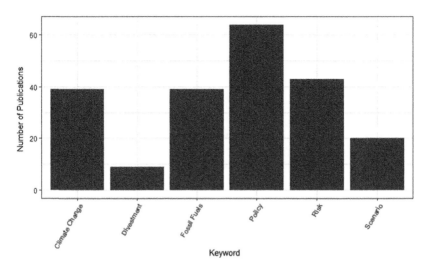

Fig. 3.1 Recent literature on stranded assets: keyword analysis (*Source* Authors' elaboration)

for example, examine how the global carbon budget has raised disagreements between opposing advocacy coalitions challenging the status quo around Norwegian petroleum resource governance, posing a risk of stranding to proven carbon reserves. Carbon risk also refers to the financial risk associated with holding carbon assets in investment portfolios. Fang et al. (2019) propose a scenario-based framework that internalises carbon risk in portfolios as a means to quantify stranded asset risk for investors. They find that markets are gradually pricing climate change risk, as attested by an inferior risk-adjusted performance of carbon-intensive industries.

Conversely, Byrd and Cooperman (2018) assert investor valuations depending on future prospects, and thus technological advancements may prove to reduce stranded asset risk. Advancements in carbon capture and sequestration development and deployment, for example, may improve the economic viability of continued coal-based energy production and thus mitigate the costs associated with stranded assets. We find that researchers are actively examining how limits to carbon emissions not only affect the direct assets that are stranded but how the associated assets may affect other stakeholders, like governments and investors. We also find that multiple drivers, like advocacy groups and technological advancements, influence the pace of which assets may be stranded.

The literature on financial risk, though limited, incorporates another essential stakeholder, long-term investors. The discourse around maligned incentives, short-termism and fiduciary duty seem to be particularly prominent in the context of stranded assets, given the time horizon over which the risks will be realised. Long-term holdings like bonds or pension funds have significant sway in how much carbon is locked-in and is also particularly sensitive to future risks of stranded assets. Buhr (2017) examines how stranded asset risk may be better managed for long-term investors. She poses that traditional environmental, social, and governance indicators fall short when assessing stranded asset risk and alternatively propose a framework that incorporates operational risk, risks related to climate mitigation and adaptation, and natural capital risks—the latter two of which are likely to be significant and irreversible. Barker et al. (2016) look to the fiduciary obligations of pension funds. They argue that passive or inactive governance of climate change risk is unlikely to satisfy the funds' fiduciary duty to their constituents. However, outdated methodologies and assumptions continue to prefer investment as usual. Thomä and Chenet (2017) consider the theoretical lens to determine the extent to which market failures result in potential mispricing of stranded asset risks. They conclude that these market failures will require policy intervention that addresses the design of financial risk models, the transparency around their results and the institutions governing risk management. Finally, Silver (2017) considers why investors remain blind to the stranded asset risk, arguing that risk is a function of divergence from peers. Thus, changes in regulation and theory will be foundational to overhaul the cultural constraints and maligned incentives that restrict incorporating stranded asset risk into the decision-making chain. Again this literature on the financial risk associated with stranded assets reinforces the multitude of confounding drivers and actors affected by and affecting stranded asset risk. The appropriate policy remains fundamental to addressing the shortcomings of conventional theories, methods and assumptions.

Several notable recent publications delve deeper into the efficacy of policy in mitigating stranded asset risk. Van der Ploeg and Rezai (2020) examine the effects of market uncertainty and how changes of climate policies affect the valuation of physical and natural capital assets. Their analytical model suggests that when effective climate policy is implemented, exploration capital and fossil fuel reserves suffer a sudden loss in value; however, a botched climate policy immediately leads to an investment boom in exploration and a surge in discoveries. Conversely, Shimbar and Ebrahimi (2020) pose that by incorporating political risk into the valuation of renewable energy investments, once non-viable projects can demonstrate attractive returns on investment. They examine this specifically in the case of political risk in developing countries and its potential for foreign direct investment. Thus, the political climate and the implementation of the low-carbon policy is essential stranded asset risk assessment.

To conclude, it is evident that there is a multitude of factors that exhibit complex interdependencies between the drivers of stranded asset risk and affected parties. We thus close with the literature on systematic risk. Keys et al. (2019) propose that stranded asset risk is an "Anthropocene risk", that emerges from human-driven processes, interacts with global social-ecological connectivity and exhibits complex cross-scale relationships. Their example on stranded assets in aquaculture demonstrates not only the range of industries exposed to stranding but the human and environmental factors driving it. They conclude that the confluence of global demand for aquaculture, along with site-specificity and weak environmental regulations, place developing regions at more risk of stranded assets.

3.4 A Typology for Identifying and Measuring Stranded Assets Risk

Our systematic review attests to the complexities involved in identifying and measuring stranded asset risk. There are multiple confounding factors, from the drivers of stranded asset risk to the actors affected by and affecting the drivers. We also find that stranded assets are not unique to carbon, especially given some economies' ubiquitous dependence on carbon-intensive production. Consequently, it is difficult for stranded asset researchers to account for all factors. Thus, in developing a robust method of measuring stranded asset risk, we identify a multitude of conditions that result in asset stranding. Based on the literature review, we propose a typology (Fig. 3.2) for identifying and managing stranded asset risk for consideration in future research agendas. Researchers may use this typology to manage stranded asset risks more comprehensively. We propose three attributes or classifiers that all forms of stranded assets entail. First, stranded assets may arise from several drivers, which include

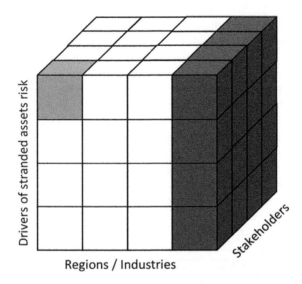

Fig. 3.2 A typology to identify and manage stranded assets risk (*Source* Authors' elaboration)

physical, transitional and regulatory factors. Second, stranded assets may differ between and across regions and industries, depending on factors like dependence on fossil fuels or sensitivity to climate change. Finally, stranded assets affect more than the devalued reserves but have systematic effects on financiers, governments and civil society.

We argue that it is important for researchers to identify how each driver might affect selected industries. This is represented by a single square in the diagram. For greater efficacy, researchers should incorporate all drivers of stranded asset risk across a region or industry. However, the most comprehensive measurement of stranded asset risk is one that incorporates costs associated not only to business operations, but financiers, politics and civil society as well. We expand on each of these attributes in more detail below.

3.4.1   Types of Stranded Assets

The cumulative limit of unburnable carbon is not the only driver of asset stranding in the low-carbon transition. Caldecott et al. (2013) present a more expansive interpretation through a typology of six environment-related risks that could cause stranded assets. Namely, assets may face physical risks associated with extreme weather events or resource degradation, as well as risks related to changing resource landscapes through the pricing and availability of substitutes. Assets may also face social risks, through government regulations, technological advancements, evolving social norms and litigation.

3.4.2 Regions and Industries

The risk of assets becoming stranded due to decarbonisation constitutes a considerable risk not only to the fossil fuel industry and other industries along its production chain but also to nations whose economies are connected to the carbon industry (Campiglio et al. 2017). For example, stranded assets can arise due to different reasons such as the development and implementation of environmental policies (Vogt-Schilb and Halle-gatte 2017), especially those geared towards climate mitigation, as well as from disruptive innovation from other industries (Green and Newman 2017). Also, other unintended consequences of fossil fuel stranding may negatively impact other sectors along the supply chain that are connected to the industry, leading to the demise or loss of assets across those sectors and other parts of the economy (Bos and Gupta 2019). Cascading effects of stranded asset risks could affect transport (Traut et al. 2018), agriculture (Marsden et al. 2019; Morel et al. 2016; Rautner et al. 2016), real estate (Muldoon-Smith and Greenhalgh 2019) and water assets (Lamb 2015). The interdependent nature of industries suggests that stranded asset risk may have systemic effects across the economy.

More, as countries intensify efforts towards economic-wide emission reduction and climate mitigation, the resultant transition to a low-carbon economy will directly impact socio-economic activities. This impact may significantly increase the risk of assets becoming stranded due to regulatory, economic or market connection to carbon-intensive industries (Bos and Gupta 2018). This will affect the productivity of carbon investment into the future, with telling effects on policies and market conditions. However, the degree of risk that individual developing and developed countries are exposed differ significantly because of diverse issues like economic situation, geographical location, environment and socio-economic capacity for mitigation and adaptation (Caldecott et al. 2016).

The determining factor into how this plays out is based on the exposure of respective countries to fossil fuel assets and resources and the concentration of the economies in specific industries. For example, countries with fossil fuel assets and resources will be at more risk of having their assets becoming stranded, and economies affected (Mercure et al. 2018). However, this may be relatively different from one country to another since the level of economic integration across regions, countries and industries differs. Likewise, coastal nations with substantial investment in tourism and investment in assets on the coastline, stand the risk of facing the effects of sea level rise and flooding that may lead to loss of those assets value or complete destruction of their economic activities. It is therefore not unusual that in the face of rising climate impact and global emission target reduction, there is the concern that any change in regulation with significant mitigating efforts will exposure the economy and assets. The exposure might be detrimental to carbon assets and resources, especially for developing and economically exposed countries (Bos and Gupta 2018; McGlade and Ekins 2015).

3.4.3   Stakeholders

Failure to deal with stranded assets will eventually damage shareholders, consumers, and overall welfare. We have seen notable action taken by many actors, towards mitigating stranded asset risk and advancing the low-carbon transition. However, some aspects remain underrepresented in research. Specifically, we look at the role of civil society, the role of industry and finally, the role of financiers and regulators.

By 2012, the 2 °C low-carbon transition and stranded assets drove climate activists to shape their strategy of targeting fossil fuel–supply financing (via divestment) and infrastructure (via protests), by casting the fossil fuel industry as "enemies". We have seen in recent years, a sense of increasing descent against carbon producers. What started from fringe movements of divestment and 'keep it in the ground' protests have now gained legitimacy among once ambivalent financiers. Today, the fossil fuel industry is perceived similarly to the tobacco industry, and industries supporting the South Africa Apartheid (Hunt et al. 2017). Divestment campaigns have directly damaged fossil fuel interests, through delaying them, cutting access to financing as they are seen as increasingly uneconomical and potentially cancelling them entirely (Carter et al. 2019).

Around the same time, the fossil fuel divestment movement urged investors to withdraw their money from the fossil fuel industry in response to the financial risks of stranded assets brought about by the carbon budget. While divestment did start with an ethical case if it is wrong to wreck the planet, then it is wrong to profit from that wreckage, there is now a clear financial case for divestment (Dordi and Weber 2019; Hunt and Weber 2018; Trinks et al. 2018). Stranded assets have underpinned arguments by civil society campaigns attempting to secure rapid economy-wide decarbonisation to reduce the scale of anthropogenic climate change. However, the costs to civil society remain under-represented in stranded asset literature (Bang and Lahn 2019; Caldecott and Dericks 2018).

Industries must also internalise stranded asset risk in corporate decision-making. They must not only act in accordance to minimise their risk, but face physical and liability risks if they fail to adequately assess the impacts of climate-related issues on corporate risk and strategy (Barker 2018). However, risks remain mispriced or left ignored because of biases, misaligned incentives and endemic short-termism (Caldecott and Dane 2015; Carney 2015; Kay 2012). These risks are exacerbated because the risks are novel, data, analytical tools and methods are missing, and there is a lack of viable options to hedge risk.

Financiers are likely to bear the risks of stranded assets, through the rapid devaluation of holdings. However, the financial sector also has the means to incentivise mitigation through lower interest rates and lower cost refinancing of green developments. The means to measure and manage the exposure of investments is still lacking. Financiers may face stranded investments through a decline in equity value and through loan defaults if they do not manage environmental risks associated with their current investments. However, financiers may also be a catalyst for new investments that incentive mitigation—through lower interest rates, loan guarantees, first loss insurance, credit risk insurance and lower cost refinancing through bonds.

Governments must adopt active regulations that encourage investments towards a low-carbon transition. Regulatory measures must increase pressure on companies, investors and civil society, but also becoming more consistent and predictable. Examples could include tax incentives, carbon markets, energy performance regulations, building standards, retrofits and concessional finance for zero-carbon technologies. Central banks and financial regulators also have an important role in maintaining financial stability (Batten et al. 2017). Clarifying fiduciary duty and addressing some of the perverse incentives of liquidity requirements would encourage long-termism. Asset owners also need to have to conviction to create long-term mandates. Stranded assets can only achieve so much if they operate in an institutional environment that is endemically short-term.

3.5   Case Study: Stranded Assets by Region

We first apply a geographical lens to regulation. Globally, countries have committed to independent nationally determined contributions (INDC) towards emission reduction targets. Thus, regions should consider how "committed emissions" should influence decarbonisation plans developed by governments (Davis et al. 2010; Davis and Socolow 2014; Pfeiffer et al. 2016). There is also a recognition that stranded asset risks span beyond the energy industry—through, for example, the physical risk of coastal properties that may be affected by extreme weather events. Thus, geographical relevance can influence optimal policy decisions.

It has been estimated that implementing the Paris Agreement will directly require that most of the known global fossil fuel reserves remain untapped. Understanding the implication of this is important since any shift towards decarbonisation globally will have a direct impact on financial markets, assets and governance (Ashford and Renda 2016). There is also an indirect impact on other investments around the fossil fuel industry, which will become high risk with grave implications for several countries and parts of the world where the economies are mainly dependent on oil output (Bos and Gupta 2019). The result of this is an unforeseen market condition or regulatory change that may lead to massive regulatory intervention while putting markets and investments at risk (Caldecott et al. 2016).

These risks cut across respective economies and industries in diverse ways, with industries such as agriculture being highly exposed due to its high dependence on fossil fuel and high emission rate across its value chain (Caldecott et al. 2013). Also, countries' interests and priorities are not the same, which makes understanding how different factors like regulation, countries' economic interests and priorities, and stakeholders pressure and inputs increase the risk of investments and assets in a specific economic sector becoming stranded.

3.5.1 Approach to Stranded Assets by Different Economies

The commitment to decarbonisation and development of Intended Nationally Determined Contributions (INDCs) for respective countries' implementation of the Paris Agreement has placed the issue of decarbonisation at the forefront of global political landscape. It has also pitted the interest of developed and often less resource-dependent economies with the developing and resource-based economies on the argument of right to development and climate justice. Developing and resource-dependent countries claim the right to develop their resources as key to national survival (Swilling and Annecke 2012).

It is important to note that while there is a lot of literature on the impact of stranded assets on the economy, little has been done with reference to its connection to developmental issues and other drivers of stranded assets. This sentiment is echoed in the level of disparity and differences between the rich developed countries when compared to resource-based and developing economies in the content and focus of their individual INDCs. This is also evident in the approach that different economies took to achieve their INDC, which is discussed in more detail below.

It should be noted that this contribution is not focused on the right to develop resources versus sustainable development impact (Bos and Gupta 2019), though both have an implication on decarbonisation. Rather, it is focused on the effect of the decisions that emanate from that right and how it may lead to exposure to climate-induced results like stranded assets. Understanding this is important as it played a role in countries' approach to the Paris Agreement and was a constant sentiment in the conditionalities for the achievement of the targets across diverse developing and resource-rich nations.

Countries' efforts to balance the need for decarbonisation and the right to develop their carbon resources constitutes a challenge for the achievement of their target climate objectives. Unfortunately, this is not a highly researched areas since most studies have been focused on developed economies and its markets, with less focus on developing economies (Caldecott et al. 2013), even though stranded assets have the potential of impacting developing economies more (Bos and Gupta 2018). This is why taking a holistic approach into countries intended resolution and path to decarbonisation through the Paris Agreement of individual countries is important to look into as it gives an idea of how countries intend to approach their decarbonisation process.

This process involves the development of policies that may lead to more assets becoming stranded, especially in carbon-intensive industries with spillover effects on industries that are environmentally linked to the sector. The risk cuts across industries with increased physical and transition risks that may lead to unintended consequences and stranding of assets due to the development of new approaches, products, markets and regulations. This will also have direct consequences on other industries that are prone to environmental impact, which may lead to asset stranding in those sectors beyond the carbon industry (Rautner et al. 2016). The effect may lead to dire financial instability (Mercure et al. 2018) and political challenges if not well managed (Vogt-Schilb and Hallegatte 2017). As these challenges are dissimilar, it is imperative to fully understand the regional, country and economic implication of stranded asset. Knowing this will allow us to navigate the challenge of stranded assets into the future and also provide a basis for assessment of regional risk, country sector priorities.

3.5.2   Case Study Review and Assessment

To understand regional and countries' exposure to environmental risk that may lead to stranded assets and resources, we will be reviewing the submitted INDC to UNFCCC by respective countries as part of their contribution to the global goal of achieving the 1.5 degrees set in the Paris Agreement. This process will also provide us with insight into the exposure of respective nations and the identified sectors at risk based on particular countries' summation and the sector at risk of direct climate impact.

By identifying this, we will have the opportunity to highlight sectors that are key areas where resources should be allocated as decarbonisation takes a foothold (Bos and Gupta 2019). This will enable us to assess countries' exposures and areas that can be affected by climate policy or commitment, such as the ones contained in the INDC. It will also provide an insight into other industries that may be affected by the reduction of carbon through mitigation efforts or have assets at risk due to climate impact that requires mitigation despite their importance to the economy.

This paper reviewed the INDC of countries that made up the G20 to understand the possible sectors of interest for climate mitigation as well as the at-risk sectors for impact based on individual countries' priorities. In addition to the G20, we reviewed developing countries across all continents and economic blocs. In all, we picked between three and five respective economies in each continent to review their respective INDCs and lastly to ensure balance we reviewed the INDCs of Small Island Developing States (SIDS) from the Pacific and the Caribbean due to their unique geographical location, low-carbon emission but high potential climate impact.

To ensure consistency, aside from the G20 other countries were picked based on their economic strength, resource base, population, or geographical spread. For example, in Africa, Nigeria and South Africa were selected based on being the two most important economies on the continent and their geographical local on the continent, with the former representing West Africa and the latter being from Southern Africa. Egypt, as the third-biggest economy, was chosen from Northern Africa, while Kenya as the biggest economy from central and east Africa, completes the list. Thus, representing all economic blocs and the geographical spread of the continent. The idea is to create both geographical, climatic and economic balance in a highly diverse continent.

Since the INDCs did not follow any pattern or expectations, the respective countries adopted different approaches to drafting their commitments. To mitigate this, we assessed individual contribution documents.

3.5.3   INCDs and Sector Breakdown

As mentioned earlier, there is no uniform approach to respective countries commitment. Hence, we had to determine countries' priorities through the relevance of key sectors and climate event as noted in the INDCs.

3.5.3.1   Global Outlook for Assets at Risk

The distinction in priorities and at-risk areas are clearly shown in the classification of risks between developing and developed countries. While developing countries are mainly sector focused in their approach and target, especially considering how climate-induced impacts may affect assets and other issues in specific sectors, the developed and fast developing countries in Asia, Europe and America tend to focus on an

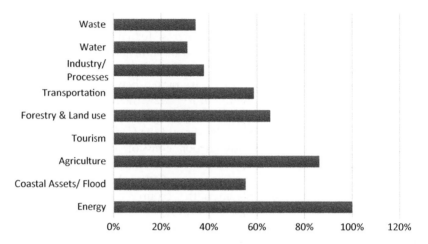

Fig. 3.3   Global assets at risk by sector (*Source* Authors' elaboration)

economy-wide approach. This may be a result of the level of development of various sectors in the economies of developed countries compared to the often resource-based economies in the less developed nations. One similarity here is that all countries perceive energy to be a major sector at risk (Fig. 3.3)

Ironically, despite concerns with energy and energy sources (including fossil fuel) and commitments made in line with this in every single INDC reviewed, the continued investment in fossil fuel industry remains unabated. The industry received over $700 billion financing for the expansion of new projects since the 2015 Paris agreement was signed. The inaction and possible financial instability challenges bear the risk ahead considering the likely impact of a transition to a low-carbon economy and how it may be fatal for carbon assets. Unfortunately, this also places assets on the oil value chain and other connected industries in the energy sector, such as power, at risk of becoming stranded (Campiglio et al. 2017).

3.5.3.2   Africa

Despite being primarily made up of developing and less economically buoyant countries, Africa has the most diverse background of all the continents reviewed with the focus mainly based on the economic realities of the nations and their geographical context (a good example is

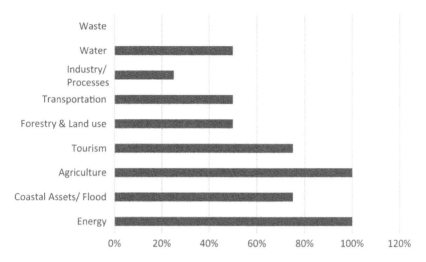

Fig. 3.4 Assets at risk by sector: Africa (*Source* Authors' elaboration)

agriculture). But when compared to issues such as water, only Egypt and South Africa felt to be water-stressed and find assets and investments in the sector to be at risk due to their geographical locations and the heavy impact of climate change on rain in South Africa and water access in the arid regions of Egypt. There is also a low focus on industrial processes, with only Nigeria highlighting industry and production as a priority and no country seeing issues with waste and assets around it as a priority. This is not surprising considering the low level of development across the continent and since most African countries rely heavily on imports from other parts of the world. Figure 3.4 depicts assets at risk by sector in the continent.

3.5.3.3   EU, Asia and North America
Europe and North America have economic plans to tackle the greenhouse gas (GHG) emissions which includes carbon dioxide (CO_2), methane (CH_4), nitrous oxide (N_2O), perfluorocarbons (PFCs), hydrofluorocarbons (HFCs), sulfur hexafluoride (SF_6) and nitrogen trifluoride (NF_3) across all economic sectors. Hence it is not surprising that while sectors like waste, which had little or no mention in African countries' plans,

were given a prominent role in the EU, at the same extent as energy, agriculture, industry, forest and land use (Fig. 3.5).

The interesting observation in North America is the position of Mexico which towed the pattern of other developing economies by highlighting concerns with coastal areas and tourism and other at-risk sectors, while placing less importance on issues of waste and industry (both being of great concern developed nations hence bringing the diversity to the perception of sectors that are prone to the impact of transition risk of climate change.

The North American reality (see Fig. 3.6) is very similar to that of Asia where the more advanced economies like China, South Korea and Japan had exactly same realities and plan towards an economic-wide effort based on identified GHG emission with major impacts expected across waste, industry/production processes, transportation and energy. However, there are differences to countries like India placing a premium on issues such as water, which is mainly due to the water stress of its northern region and places like Indonesia that has a substantial length of its coastal area assets at risk of flooding and investments being submerged by rising sea levels (see also Fig. 3.7).

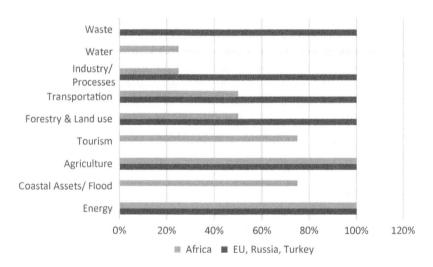

Fig. 3.5   Assets at risk by sector: Africa vs cluster EU, Russia and Turkey (*Source* Authors' elaboration)

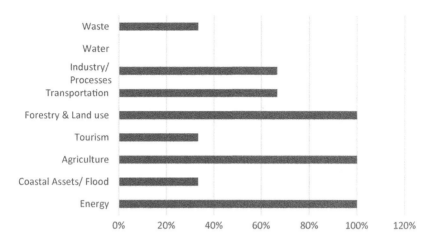

Fig. 3.6 Assets at risk by sector: North America (*Source* Authors' elaboration)

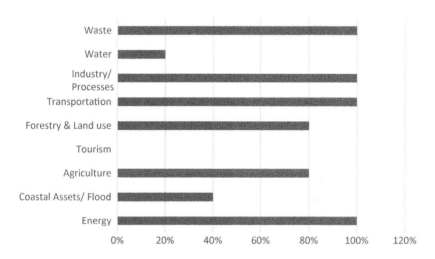

Fig. 3.7 Assets at risk by sector: Asia (*Source* Authors' elaboration)

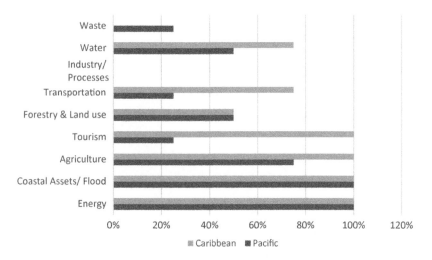

Fig. 3.8 Assets at risk by sector: Pacific and Caribbean countries (*Source*: Authors' elaboration)

3.5.3.4   SIDS—Caribbean and Pacific

The SIDS countries' approach in the INDC and their classification of sectors are similar. This could be a result of their geographical location and their vulnerability as small countries with little or almost inconsequential contribution to climate impact but being more vulnerable to its impact with a high possibility of leaving several assets in the countries stranded and economies negatively impacted. It should be noted that there are major similarities between Caribbean nations and Pacific countries on exposures in several areas, especially on issues like the priority of coastal assets and investment. Unfortunately, these risks are linked to the tourism industry in most countries especially in the Caribbean, where all the countries depend primarily on tourism for economic growth (Fig. 3.8). Because of the difference in geographical and economic arrangements the Caribbean countries seem to have a different level of risk.

3.5.3.5   Middle East

The Middle East (Fig. 3.9) gives an interesting angle of observation for stranded assets. The region is rich in fossil fuel, and the economies of the countries rely on the industry. In fact, in most cases, it is the principal source of government income. In addition, the arid-climate conditions of

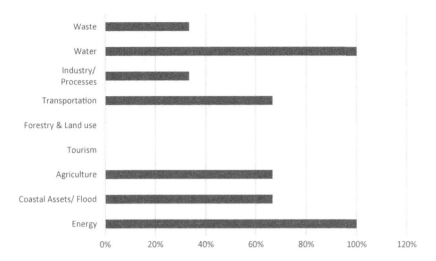

Fig. 3.9 Assets at risk by sector: Middle East (*Source* Authors' elaboration)

the region imply that vast parts of land are water-stressed, and new technology needs to be developed for industrial and domestic water needs. This has led most countries to invest in the desalination of seawater due to a shortage of groundwater and other freshwater sources. It is not surprising that these two issues constitute the most critical issues among the countries reviewed in the region. So, while there are conflicting approaches to other essential areas in terms of priorities and economic arrangement, the three countries assessed together, Saudi Arabia, Kuwait and United Arab Emirate (UAE), seem to be focused on mitigating climate impacts, but without any concrete commitment to reducing their oil production output. Rather, there are references to increasing technology to create "cleaner crude", but with no direct obligation to reduce or stop fossil fuel exploration, which in most cases is perceived as key to the region's economic prosperity.

3.6   A Research Agenda for Stranded Assets Risk

One of the key observations of this study is how the approach to stranded asset has overtly focused on carbon assets over the years while negating the roles played by other factors that increase stranded asset risks. Key among this are the different drivers stemming from stakeholder interests and priorities. A cursory look into the INDCs shows a consistent approach in the documents wherein most were carefully crafted, often not necessarily to protect physical existing assets, but interests. A lot of the commitments stems from attempts by policymakers to try to ameliorate challenges being faced in specific sectors that are prone to risks, while seeking alternatives in investment in technologies and other areas that may enable a smooth transition despite obvious high-risk levels for the sectors that are prioritised.

This observation in the reviewed INDCs directly aligns with the earlier discussed typology for this study as there is evidence of diverse drivers of stranded asset risks that most literature is not taking into full consideration. Starting with regulatory drivers, one outstanding and recurring factor sits with existing climate action and policies in individual countries upon which a couple of the INDCs commitments were based. Over 75% of the INDCs reviewed referred to a form of the existing regulatory framework, climate plan or environmental policy and drivers that provides a basis for national guidance on industries and other areas of national interests. These highlighted areas, some of which are based on national or regional interests, often set goals and climate targets for specific areas of interventions and where to prioritise climate mitigation and adaptation.

It was further observed that these prioritised areas are based on issues of interest and regional economic dependencies. For example, none of the Middle East oil-producing countries developed their INDCs without a provision for fossil fuel in their future, which are often the lifeline of their economies, in their future. Though countries in these regions noted the need for economic diversification and GHG emission reduction, there were no firm commitments to achieve this through a reduction in fossil fuel exploration. Still, they generally proposed a more adaptive approach using technology such as carbon capture (Saudi Arabia), production of clean fuel (Kuwait) and utilisation of technologies to improve efficiency and reduce emissions (UAE). This seems to assert that factors such as economic survival, regional or national interest, and other perceived priorities play more prominent roles in decision-making and dwarfs the risk of stranded assets in a high-risk industry, especially at the local and national levels.

The approach of individual national and regional interests is in line with the fact that stakeholders and industries are often buoyed by regulators and political class concerns about the systematic effects of decisions on key interest areas. Hence, stakeholders and other interest groups consistently play important roles in the implementation of key policies that eventually help in the formulation of regulations and processes. A good example here is the issue of job provision, which is constantly referred to in the INDCs, especially of most developing economies. To protect jobs, the political class will often promote or protect interests in industries that will enhance issues of economic well-being even if such have a lasting effect and environmental impact on the long run. This position can be exacerbated by interest and pressure groups that may slow the transition or protect their respective interests either for or against specific sectors regardless of how that industry is being perceived.

Research is nascent, with a focus on identifying how selected aspects of stranded assets may play out. However, to better manage the financial risk, a broader perspective that incorporates the different types of stranded asset risk is necessary. Our typology proposes a means by which future researchers can systematically examine how to evaluate the risk of stranded assets. Our case study of INDC reports attests that regions around the world face various and unique risks of stranded assets based on their materiality. However, the risks remain within the realm of stranded investments for the region or industry and consequently fail to consider the effects of stranded assets on investors, governments and civil society. Thus, future research on stranded assets would benefit from encompassing the proposed typology, to gain a comprehensive understanding of stranded asset risk.

Our case studies indicate that research into the risk of stranded assets needs to be expanded beyond assets and machinery. There is a need to investigate what other drivers may contribute to the risk of stranded assets, particularly in specific sectors and regions, and how related industries maybe be able to mitigate the risks. Issues such as political interests that relates to impacts on the economy and jobs have also been found to fuel the political will to act in specific areas. There is, therefore a need to understand and study, for example, what still drives the continued reliance and investment in fossil fuel and carbon industries beyond the environmental argument. Part of this may involve recognising and addressing stakeholders' concerns on livelihood and economic impact, which will help drive economic and political cooperation.

REFERENCES

Allen, M. R., Frame, D. J., Huntingford, C., Jones, C. D., Lowe, J. A., Mein-shausen, M., et al. (2009). Warming caused by cumulative carbon emissions towards the trillionth tonne. *Nature, 458*(7242), 1163–1166. https://doi.org/10.1038/nature08019.

Ansar, A., Caldecott, B., & Tilbury, J. (2013). *Stranded assets and the fossil fuel divestment campaign: What does divestment mean for the valuation of fossil fuel assets*. Stranded Assets Programme, SSEE, University of Oxford.

Arbuthnott, K. D., & Dolter, B. (2013). Escalation of commitment to fossil fuels. *Ecological Economics, 89*, 7–13. https://doi.org/10.1016/J.ECOLECON.2013.02.004.

Ashford, N. A., & Renda, A. (2016). Aligning policies for low-carbon systemic innovation in Europe (September 27, 2016). *CEPS special report, 2016.* Available at SSRN: https://ssrn.com/abstract=2859424.

Bang, G., & Lahn, B. (2019). From oil as welfare to oil as risk? Norwegian petroleum resource governance and climate policy. *Climate Policy.* https://doi.org/10.1080/14693062.2019.1692774.

Barker, S. (2018). An introduction to directors duties in relation to stranded asset risk. In B. Caldecott (Ed.), *Stranded assets and the environment: Risk, resilience and opportunity* (pp. 221–271). Abingdon: Routledge.

Barker, S., Baker-Jones, M., Barton, E., & Fagan, E. (2016). Climate change and the fiduciary duties of pension fund trustees—Lessons from the Australian law. *Journal of Sustainable Finance and Investment.* https://doi.org/10.1080/20430795.2016.1204687.

Batten, S., Sowerbutts, R., & Tanaka, M. (2017). Let's talk about the weather: The impact of climate change on central banks. *SSRN Electronic Journal.* https://doi.org/10.2139/ssrn.2783753.

Bos, K., & Gupta, J. (2018). Climate change: The risks of stranded fossil fuel assets and resources to the developing world. *Third World Quarterly.* https://doi.org/10.1080/01436597.2017.1387477.

Bos, K., & Gupta, J. (2019). Stranded assets and stranded resources: Implications for climate change mitigation and global sustainable development. *Energy Research and Social Science.* https://doi.org/10.1016/j.erss.2019.05.025.

Buhr, B. (2017). Assessing the sources of stranded asset risk: A proposed framework. *Journal of Sustainable Finance and Investment, 7*(1), 37–53. https://doi.org/10.1080/20430795.2016.1194686.

Byrd, J., & Cooperman, E. S. (2018). Investors and stranded asset risk: Evidence from shareholder responses to carbon capture and sequestration (CCS) events. *Journal of Sustainable Finance and Investment.* https://doi.org/10.1080/20430795.2017.1418063.

Caldecott, B., & Dane, R. (2015). *Investment consultants and green investment: Risking stranded advice?*

Caldecott, B., & Dericks, G. (2018). Empirical calibration of climate policy using corporate solvency: A case study of the UK's carbon price support. *Climate Policy*. https://doi.org/10.1080/14693062.2017.1382318.

Caldecott, B., Harnett, E., Cojoianu, T., Kok, I., & Pfeiffer, A. (2016). *Stranded assets: A climate risk challenge*. Inter-American Development Bank.

Caldecott, B., Howarth, N., & McSharry, P. (2013). *Stranded assets in agriculture: Protecting value from environment-related risks*. Stranded Assets Programme, SSEE, University of Oxford.

Campiglio, E., Godin, A., Kemp-Benedict, E., & Matikainen, S. (2017). The tightening links between financial systems and the low-carbon transition. *Economic Policies Since the Global Financial Crisis*. https://doi.org/10.1007/978-3-319-60459-6_8.

Carney, M. (2015). *Breaking the tragedy of the horizon—Climate change and financial stability*. Speech given at Lloyd's of London by the Governor of the Bank of England, 29.

Carter, A. V., McKenzie, J., & Salam, J. (2019). *Amplifying 'keep it in the ground' first-movers: Toward a comparative framework*. Congress of the Humanities and Social Sciences.

Davis, S. J., Caldeira, K., & Matthews, H. D. (2010). Future CO_2 emissions and climate change from existing energy infrastructure. *Science*. https://doi.org/10.1126/science.1188566.

Davis, S. J., & Socolow, R. H. (2014). Commitment accounting of CO_2 emissions. *Environmental Research Letters*. https://doi.org/10.1088/1748-9326/9/8/084018.

Dordi, T., & Weber, O. (2019). The impact of divestment announcements on the share price of fossil fuel stocks. *Sustainability, 11*(11), 3122. https://doi.org/10.3390/su11113122.

Fang, M., Tan, K. S., & Wirjanto, T. S. (2019). Sustainable portfolio management under climate change. *Journal of Sustainable Finance and Investment*. https://doi.org/10.1080/20430795.2018.1522583.

Green, J., & Newman, P. (2017). Disruptive innovation, stranded assets and forecasting: The rise and rise of renewable energy. *Journal of Sustainable Finance & Investment, 7*(2), 169–187. https://doi.org/10.1080/20430795.2016.1265410.

Heede, R. (2014). Tracing anthropogenic carbon dioxide and methane emissions to fossil fuel and cement producers, 1854–2010. *Climatic Change, 122*(1–2), 229–241. https://doi.org/10.1007/s10584-013-0986-y.

Hunt, C., & Weber, O. (2018). Fossil fuel divestment strategies: Financial and carbon related consequences. *Organization & Environment, 32*, 41–61.

Hunt, C., Weber, O., & Dordi, T. (2017). A comparative analysis of the anti-Apartheid and fossil fuel divestment campaigns. *Journal of Sustainable Finance and Investment, 7*(1). https://doi.org/10.1080/20430795.2016.1202641.

Jaccard, M., Hoffele, J., & Jaccard, T. (2018). Global carbon budgets and the viability of new fossil fuel projects. *Climatic Change, 150*(1–2), 15–28. https://doi.org/10.1007/s10584-018-2206-2.

Kay, J. (2012, July). *The Kay review of UK equity markets and long-term decision making* (Final Report).

Keys, P. W., Galaz, V., Dyer, M., Matthews, N., Folke, C., Nyström, M., et al. (2019). Anthropocene risk. *Nature Sustainability*. https://doi.org/10.1038/s41893-019-0327-x.

Knuth, S. (2017). Green devaluation: Disruption, divestment, and decommodification for a green economy. *Capitalism, Nature, Socialism*. https://doi.org/10.1080/10455752.2016.1266001.

Lamb, C. (2015). *Drying and drowning assets—How worsening water security is stranding assets*. http://www.strandedassets2015.org/agenda.html; http://www.strandedassets2015.org/uploads/2/6/9/5/26954337/session_v_presenter_ii_catelamb.pdf.

Marsden, T., Moragues Faus, A., & Sonnino, R. (2019). Reproducing vulnerabilities in agri-food systems: Tracing the links between governance, financialization, and vulnerability in Europe post 2007–2008. *Journal of Agrarian Change*. https://doi.org/10.1111/joac.12267.

McGlade, C., & Ekins, P. (2015). The geographical distribution of fossil fuels unused when limiting global warming to 2 °C. *Nature*. http://www.nature.com/nature/journal/v517/n7533/abs/nature14016.html.

Meinshausen, M., Meinshausen, N., Hare, W., Raper, S. C. B., Frieler, K., Knutti, R., et al. (2009). Greenhouse-gas emission targets for limiting global warming to 2 °C. *Nature, 458*(7242), 1158–1162. https://doi.org/10.1038/nature08017.

Mercure, J.-F., Pollitt, H., Viñuales, J. E., Edwards, N. R., Holden, P. B., Chewpreecha, U., et al. (2018). Macroeconomic impact of stranded fossil fuel assets. *Nature Climate Change*. https://doi.org/10.1038/s41558-018-0182-1.

Meyer, K. R. (1997). Restructuring and the market-to-book ratio. *Electricity Journal*. https://doi.org/10.1016/S1040-6190(97)80349-1.

Morel, A., Friedman, R., Tulloch, D. J., & Caldecott, B. (2016). *Stranded assets in palm oil production: A case study of Indonesia about the sustainable finance programme*. Smith School of Enterprise and the Environment, University of Oxford.

Muldoon-Smith, K., & Greenhalgh, P. (2019). Suspect foundations: Developing an understanding of climate-related stranded assets in the global real estate sector. *Energy Research and Social Science*. https://doi.org/10.1016/j.erss.2019.03.013.

Odenberger, M., & Johnsson, F. (2007). Achieving 60% CO_2 reductions within the UK energy system-Implications for the electricity generation sector. *Energy Policy*. https://doi.org/10.1016/j.enpol.2006.08.018.

Odenberger, M., Unger, T., & Johnsson, F. (2009). Pathways for the North European electricity supply. *Energy Policy*. https://doi.org/10.1016/j.enpol.2008.12.029.

Pearson, R. J., Turner, J. W. G., Eisaman, M. D., & Littau, K. A. (2009). Extending the supply of alcohol fuels for energy security and carbon reduction. *SAE Technical Papers*. https://doi.org/10.4271/2009-01-2764.

Perez, C. (1985). Microelectronics, long waves and world structural change: New perspectives for developing countries. *World Development*. https://doi.org/10.1016/0305-750X(85)90140-8.

Perez, C. (2009). Technological revolutions and techno-economic paradigms. *Cambridge Journal of Economics*. https://doi.org/10.1093/cje/bep051.

Pfeiffer, A., Millar, R., Hepburn, C., & Beinhocker, E. (2016). The '2°C capital stock' for electricity generation: Committed cumulative carbon emissions from the electricity generation sector and the transition to a green economy. *Applied Energy*. https://doi.org/10.1016/j.apenergy.2016.02.093.

Quéré, C. Le, Andres, R. J., Boden, T., Conway, T., Houghton, R. A., House, J. I., et al. (2013). The global carbon budget 1959–2011. *Earth System Science Data*, 5(1), 165–185.

Randalls, S. (2010). History of the 2°C climate target. *Wiley Interdisciplinary Reviews: Climate Change*, 1(4), 598–605. https://doi.org/10.1002/wcc.62.

Rautner, M., Tomlinson, S., & Hoare, A. (2016). *Managing the risk of stranded assets in agriculture and forestry* (Chatham House Research Paper).

Richels, R. G., Blanford, G. J., & Rutherford, T. F. (2009). International climate policy: A "second best" solution for a "second best" world? *Climatic Change*. https://doi.org/10.1007/s10584-009-9730-z.

Rogelj, J., Shindell, D., Jiang, K., Fifita, S., Forster, P., Ginzburg, V., et al. (2018). *Mitigation pathways compatible with 1.5 °C in the context of sustainable development—Global warming of 1.5 °C—An IPCC special report on the impacts of global warming of 1.5 °C above pre-industrial levels and related global greenhouse gas emission pathways, in the context of strengthening the global response to the threat of climate change*. https://doi.org/10.1017/CBO9781107415324.

Rothkopf, M. H. (1997). On misusing auctions to value stranded assets. *Electricity Journal*. https://doi.org/10.1016/S1040-6190(97)80315-6.

Schumpeter, J. A. (1942). Capitalism and the process of creative destruction. *Monopoly Power and Economic Performance*. https://doi.org/10.1017/CBO9781107415324.004.

Shimbar, A., & Ebrahimi, S. B. (2020). Political risk and valuation of renewable energy investments in developing countries. *Renewable Energy*. https://doi. org/10.1016/j.renene.2019.06.055.

Silver, N. (2017). Blindness to risk: Why institutional investors ignore the risk of stranded assets. *Journal of Sustainable Finance and Investment*. https://doi. org/10.1080/20430795.2016.1207996.

Stern, J. (2012). Developing upstream competition in the England and Wales water supply industry: A new approach. *Utilities Policy*. https://doi.org/10. 1016/j.jup.2011.11.007.

Strauch, Y., Dordi, T., & Carter, A. V. (2020). Constraining fossil fuels based on 2 °C carbon budgets: The rapid adoption of a transformative concept in politics and finance. *Climatic Change, 160,* 181–201.

Swilling, M., & Annecke, E. (2012). *Just transitions: Explorations of sustainability in an unfair world*. United Nations University Press.

Thomä, J., & Chenet, H. (2017). Transition risks and market failure: A theoretical discourse on why financial models and economic agents may misprice risk related to the transition to a low-carbon economy. *Journal of Sustainable Finance and Investment, 7*(1). https://doi.org/10.1080/20430795.2016. 1204847.

Traut, M., Larkin, A., Anderson, K., McGlade, C., Sharmina, M., & Smith, T. (2018). CO_2 abatement goals for international shipping. *Climate Policy*. https://doi.org/10.1080/14693062.2018.1461059.

Trinks, A., Scholtens, B., Mulder, M., & Dam, L. (2018). Fossil fuel divestment and portfolio performance. *Ecological Economics, 146*(1), 740–748.

van der Ploeg, F., & Rezai, A. (2020). The risk of policy tipping and stranded carbon assets. *Journal of Environmental Economics and Management*. https://doi.org/10.1016/j.jeem.2019.102258.

Vogt-Schilb, A., & Hallegatte, S. (2017). Climate policies and nationally determined contributions: Reconciling the needed ambition with the political economy. *Energy and Environment*. https://doi.org/10.1002/wene.256.

Woychik, E. C. (1995). A California restructuring settlement: Will all the parties agree? *The Electricity Journal*. https://doi.org/10.1016/1040-6190(95)901 05-1.

CHAPTER 4

Sustainability-Related Risks, Risk Management Frameworks and Non-financial Disclosure

Marco Migliorelli and Vladimiro Marini

Abstract This chapter gives an overview of the main strategic and organisational implications for financial institutions when fully considering the actual and potential impacts of sustainability-related risks on their businesses. In this respect, the chapter first argues that, to ensure the effectiveness of the general risk management framework, developments

---

The contents included in this chapter do not necessarily reflect the official opinion of the European Commission. Responsibility for the information and views expressed lies entirely with the authors.

M. Migliorelli (✉)
IAE Université Paris 1 Panthéon-Sorbonne (Sorbonne Business School), Paris, France
e-mail: Marco.Migliorelli@ec.europa.eu

European Commission, Brussels, Belgium

V. Marini
University of Rome Tor Vergata, Rome, Italy

© The Author(s) 2020
M. Migliorelli and P. Dessertine (eds.), *Sustainability and Financial Risks*, Palgrave Studies in Impact Finance,
https://doi.org/10.1007/978-3-030-54530-7_4

are necessary at several levels of the organisation, in particular within the perimeter of competence of the management board, the risk management function and the operational business units. Then, the chapter discusses the issue of disclosing sustainability-related information by illustrating existing industry and policy standards. It concludes that more work is still needed in terms of quality and comparability of the information to foster market discipline via the disclosure of sustainability-related information.

Keywords Sustainability-related risks · Climate change-related risks · Risk management frameworks · Non-financial disclosure · Sustainability disclosure · Sustainable finance

4.1   INTRODUCTION

The growing awareness of the materiality of sustainability-related risks[1] for the financial industry will likely produce in the next a few years a significant change in the way financial institutions deal with this issue. In the (rare) cases in which structured approaches have been already adopted, banks and other financial actors have started with considering in particular climate-related risks, by treating them as a new and relevant source of financial risk. In this respect, these organisations have put in place actions at the level of both overall strategy and risk analysis. Climate-related risks are systematically monitored and reviews of sectoral portfolios performed to quantify the level of exposure, with specific risk committees holding regular meetings dedicated to this new source of risks (ACPR 2019). In parallel, some institutions are also taking into consideration the objective of reducing the carbon footprint of their credit portfolios when designing strategic orientations, trying to aligning their funding to the 2 °C scenario

---

[1] Examples of sustainability-related risks linked to climate changes are: increase in the frequency and magnitude of floods, droughts and storms; permanent change in climate conditions; increase in the level of seas; distress and high volatility in commodity markets. Sustainability-related risks linked to environmental degradation include, but are not limited to: air or waters pollution; deforestation; loss of biodiversity. Types of sustainability-related risks linked to social inequality are: unfair treatment of workers; discriminatory treatment of women; discriminatory treatment of minorities; social dumping. From these risks direct and indirect financial risks can emerge for financial institutions. See Chapter 1 for a structured analysis.

as resulted from the Paris Agreement (e.g. ACPR 2019). Nevertheless, awareness and practices remain highly heterogeneous across the financial sector, with significant differences in the maturity and the depth of the needed organisational shift. As a matter of fact, to ensure adequate preparedness, an increasing involvement of governance bodies at the highest level of decision, the acquisition of development of internal expertise, a progressive integration of sustainability-related risks into existing risk management practices and the consequent advances in dedicated tools are all nested elements that will require increasingly attention.

Against this background, this chapter aims at giving an overview of the main strategic and organisational implications for financial institutions when fully considering the actual and potential impacts of sustainability-related risks on their businesses. In this respect, the chapter analyses, in Sect. 4.2, the main elements of the risk management framework that require a specific development, in particular within the perimeter of competence of the management board, the risk management function and the operational business units. Hence, in Sect. 4.3, the chapter discusses the issue of disclosing sustainability-related information, by illustrating existing standards and assessing the effectiveness of market discipline in the actual policy context.

4.2 Risk Management and Sustainability-Related Risks

4.2.1 Including Sustainability-Related Risks in the Risk Management Framework

Risk management is a core activity for any financial institution. Its main purpose is to ensure that all the organisation's significant risks are detected, measured, managed and reported correctly. For this reason, the risk management function is involved in the definition of the financial intermediary's risk strategy as well as in all the other decisions that have a significant influence on the level of risk the institution is ready or able to accept, in particular by providing the full picture at any point in time of the whole range of the actual and potential risks that feature the business.[2]

---

[2] Under the point of view of the wider internal control system, risk management is one of the functions that form the "three lines of defence" model adopted in the financial sector and endorsed by relevant regulation. In this respect, the company's first line of

When it comes to the assessment of the possible impacts of sustainability-related risks on financial institutions, the risk management function represents the core of the organisation's capacity to appraise the magnitude of the changes needed to the established processes and practices. Nevertheless, the role of the management board and of the business units (front office, branches and other risk-taking entities) is also relevant, as these actors are responsible for respectively deciding and implementing the risk-taking strategy of the organisation. To start portraying the possible strategic and organisational impacts of sustainability-related risks on financial institutions, Fig. 4.1 outlines a general risk management framework as typically in place in the financial industry. It is composed by the four macro phases of identification of the risk context, assessment of the risks, set-up of the risk-taking strategy and implementation of the risk-taking strategy. Within the framework, the activities that are expected to need some form of adjustment when taking into account sustainability-related risks are also highlighted (in *italic*).

4.2.2   Identifying the Risk Context

The risk context (often also referred as to risk inventory) should be considered the general infrastructure on which the governance of the risks is built. For this reason, its identification lays within the core responsibilities of the risk management function. The definition of the risk context includes the throughout analysis of the relevant categories of risks potentially having economic or financial impact on the financial institution, as stemming from the specific characteristics featuring the business and the risk-taking strategy adopted by the organisation. Categories of risks are mainly classified business-wise (e.g. credit risk, market risk, liquidity risk, operational risk) and are associated to specific risk events (e.g. for credit risk, the default of a client or the reduction in the value of the collaterals covering a loan). Ad hoc risk taxonomies and risk registers are widely used

defence is typically formed by the business units responsible for identifying the risks associated with each transaction and ensuring compliance with the established procedures and limits when dealing with these risks. The second line of defence includes risk management, compliance and (for insurance companies) the actuarial function, which collectively have to guarantee that the risks are identified and treated in accordance with the established rules. Finally, the third line of defence is formed by the internal audit, which in turn assesses alignment with rules and procedures by all actors within the company and verify the effectiveness of the internal control system.

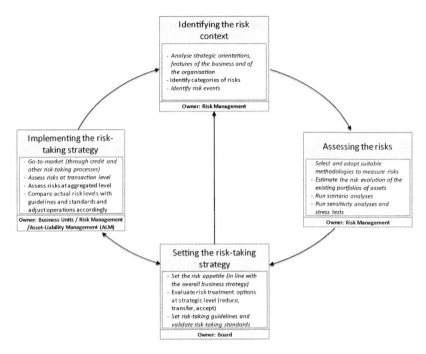

Fig. 4.1 Risk management framework and impact of sustainability-related risks (*Notes* In *italics* the activities that are impacted by the consideration of sustainability-related risks. *Source* Authors' elaboration)

by risk managers to systematically characterise the relationship between categories of risks and risk events. In identifying categories of risk and related risk events, particular attention is given to the analysis of the key areas of operations of the organisation, which are usually represented by specific economic sectors, geographies, segment of clientele, go-to-market channels, and also include an internal dimension to account for the risks linked to the functioning of the operational processes.

The integration of sustainability-related risks in the risk context is a key task for risk managers assessing the impact of sustainability on financial institutions. In this respect, it may be argued that, under a risk management perspective, sustainability-related risks do not represent a self-standing category of risks. More coherently, they represent emerging primary factors potentially able to trigger (or increase the potential negative impacts of) already considered risk events.[3] As an example, the occurrence of extreme weather events due to climate change, such as floods or hurricanes, may cause for banks' clients the disruption of the production processes or unexpected losses of assets. This may harm their solvency, and such a possible outcome should be treated by banks as an increase in the exposure to credit risk. Similarly, the same extreme weather events may produce for insurance companies unexpected higher payments on previously insured risks. This implies the need to properly manage liquidity risk and physical risks.[4]

However, even though sustainability-related risks are not expected to represent a self-standing risk category, financial institutions should systematically analyse the impacts on their business stemming from climate change, environmental degradation, social inequality and other sustainability-related factors. In particular, the possible transmission channels of sustainability-related risks on financial risks should be detected at single organisation level and integrated in the organisation's risk context. In doing that, an addition to the existing risk taxonomies and risk registers may be also foreseen.[5]

4.2.3   Assessing the Risks (Definition of Risk Scenarios)

A second macro phase featuring the risk management framework of financial institutions is the assessment of the materiality of the risks. To this extent, risk assessment is backed by the establishment and periodic update of specific methodologies, typically set at level of category of risk (in order to take into consideration the inherent features of the underlying risk events).[6] A high degree of innovation is today required as concerns

---

[3] See for example PRA (2018) or ACPR (2019).

[4] See Chapter 1 for a wider discussion on why sustainability-related risks should not be considered a new self-standing risk category.

[5] See Chapter 1 for a wider discussion and for a structured attempt to systematically link sustainability-related risks and financial risks.

[6] In this respect, abundant evidence exists dealing with the risk management methodologies to deal with the traditional categories of risks (such as market risk, credit risk or operational risk). See for example Bessis (2011), Jorion (2007), De Servigny and Renault (2004) or Christoffersen (2003).

the methodologies for the analysis of the materiality of sustainability-related risks. To this end, scenario analyses represent a central family of suitable instruments, thus far primarily developed for the assessment of the impact of climate change and the gone-with physical and transition risks[7] (e.g. Oliver Wyman 2019; PIK 2014). In general terms, scenario analyses are run to examine the financial institution's response to events or groups of events and, from there, to create information on its capacity to withstand possible shocks. Given the inherent characteristics of sustainability-related risks, which are expected to spread increasingly their effects in the long-term, scenarios should mirror plausible future climate, environmental, societal or policy developments and make use of projections factoring in the most accurate data possible at the time of the analysis. These projections should in particular be consistent with existing international agreements and initiatives, notably the Paris Agreement and the Sustainable Development Goals.[8]

In practice, building sustainability scenario analyses for financial risks assessment consists of four steps. First, the identification of a long-term status or objectives (e.g. in the case of climate change, an increase in the global average temperature of 2 °C above pre-industrial levels—in line with the upper limit of the Paris Agreement—or carbon neutrality by 2050). Second, the definition of the possible patterns to reach the identified status or objectives (e.g. to attain carbon neutrality, full electrification of the economic system or high levels of deployment of low-carbon energies). Third, the identification of the sustainability-related risks in each pattern (e.g. in case of carbon neutrality reached through high levels of deployment of low-carbon energies, the loss in the assets value of oil industries and related transition risk or—still—the occurrence of extreme

---

[7] Physical risk can be defined as the impacts today on insurance liabilities and the value of financial assets that arise from climate and weather-related events that may damage property or disrupt trade. Transition risk can be defined as the financial risk that could result from the process of adjustment towards a low-carbon economy, such as changes in policy, technology and physical risks that could prompt a reassessment of the value of a large range of assets as costs and opportunities become apparent (e.g. BoE 2015).

[8] For example, with the Paris Agreement the international community pledged for "holding the increase in the global average temperature to well below 2°C above pre-industrial levels and pursuing efforts to limit the temperature increase to 1.5°C". Under a risk management perspective, scenarios with 2 °C and 1.5 °C temperature increase can be built to assess the impact on the different economic sectors in terms of transition risk and physical risk. See, for some reference scenarios, EC (2018).

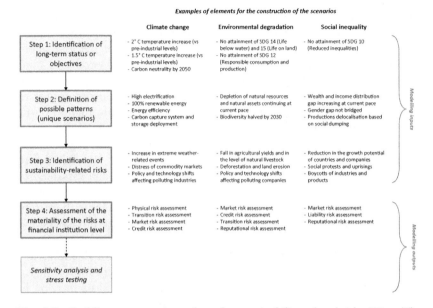

Fig. 4.2 Building-up scenario analyses for sustainability-related risks (*Notes* The list of elements for each step is illustrative and not meant to be exhaustive. *Source* Authors' elaboration)

weather events and related physical risk). Fourth, the assessment of the materiality of the identified risks for the specific financial institution (e.g. the amount of the reduction in the value of the shares of oil companies in the trading portfolios or the reduction of creditworthiness of existing and prospective clients hit by extreme weather events). Statistical models, featured by different levels of sophistication, are typically employed to support the construction of the scenarios. Figure 4.2 summarises the main steps for the construction of scenario analyses and lists a number of key parameters when specifically considering climate change, environmental degradation and social inequality[9] as sustainability-related factors.

Clearly, scenario analyses should be built on the specific characteristics of each financial institution, in particular as concerns the exposure

[9] For some additional reference on the parameters used to build climate change scenarios, see TCFD (2019, pp. 69–71).

to specific sectors, geographies and policy contexts. Each organisation should hence define the specific assumptions for the analysis based on its own (risk) profile and individual specifications, and consider several alternative possible scenarios. The outcomes of the scenario analysis may be interpreted under a quantitative or qualitative point of view, also considering the robustness of the statistical modelling supporting the analysis. Nevertheless, it is important that the outcomes consider possible short, medium and long-term impacts to feed efficiently strategic decision-making. The scenarios built can also be used to run sensitivity analyses (by foreseeing variances in critical variables) and stress testing (by factoring in the long-term capital and liquidity positions of the financial institution),[10] in this way producing information on the resilience of the organisation to sustainability-related risks.

In recent years, some concrete advances have been made in particular as concerns the construction of scenarios able to tackle the risks coming from climate change. In this respect, examples of *impact scenarios* and *transition scenarios* are progressively getting available.[11,12] These

---

[10] In this respect, it is important to underline that a number of national and international organisations, including the Network for Greening the Financial System (NGFS), the European Systemic Risk Board (ESRB) and the European Central Bank (ECB) are currently working on scenarios for climate-related stress tests.

[11] In April 2020, the Banque de France published a paper providing a tool to build climate change scenarios to forecast Gross Domestic Product (GDP), modelling both GDP damage due to climate change and the GDP impact of mitigating measures. It adopts a supply-side, long-term view, with 2060 and 2100 horizons. It is a global projection tool (30 countries/regions), with assumptions and results both at the world and the country/regional level. Five different types of energy inputs are taken into account according to their CO_2 emission factors. Full calibration is possible at each stage, with estimated or literature-based default parameters. In particular, Total Factor Productivity (TFP), which is a major source of uncertainty on future growth and hence on CO_2 emissions, is endogenously determined, with a model encompassing energy prices, investment prices, education, structural reforms and decreasing return to the employment rate. Four scenarios are also presented: Business As Usual (BAU), with stable energy prices relative to GDP price; Decrease of Renewable Energy relative Price (DREP), with the relative price of non-CO_2 emitting electricity decreasing by 2% a year; Low-Carbon Tax (LCT) scenario with CO_2 emitting energy relative prices increasing by 1% per year; High-Carbon Tax (HCT) scenario with CO_2 emitting energy relative prices increasing by 3% per year. For full information, see Banque de France (2020).

[12] In a survey delivered by the Task Force on Climate-related Financial Disclosures (TCFD) to 198 financial institutions and other organisations, 43% of respondents confirmed the use of scenario for transition risk, 33% for physical risk, 15% for other risks, 19% were developing their first scenarios and 22% did not used any scenario (due

scenarios are usually supported by complex models and have been pioneered by international organisations (e.g. IEA 2019; IPCC 2018). The specific aim of *impact scenarios* is to improve the risk management comprehension of the direct consequences of climate change on people and the environment, and from there on the financial institutions. To do that, this type of scenarios typically identify projections for economic areas and ranges of magnitude for the impact of events linked to climate change in the long-term (through 2050 or beyond), covering in particular key sectors such as agriculture, public infrastructures, manufacturing industry and energy production. World temperature increases (and related scientific literature) can also be used as key orientation, featuring the so-called *temperature scenarios*. The focus of these scenarios remains the assessment of the impacts of physical risks and, to this extent, they may employ forecasts based on statistical probabilities.[13] On the other hand, *transition scenarios* aim at increasing the understanding of the risks linked to sectors that may come under pressure as a result of policy shift and technology advances needed to reach a low-carbon economy. For this reason, these scenarios mainly model the future development of fossil fuels industries (and their relations with the rest of the market), while trying to describe coherent trajectories for the achievement of specific climate goals. To do that, relevant costs and expenditures profiles are typically first identified at a company level, by considering parameters such as the costs for carbon-emission rights or the variance in the cost of debt. In this way,

to the lack of the availability of standard scenarios and assumptions, high complexity and costs, the consideration of climate-related risks as being non-material, the focus on other priorities, the use of other methods). The respondents also highlighted the main issues encountered in developing scenarios, namely: lack of appropriately granular, business-relevant data and tools supporting scenario analysis; difficulty in determining scenarios, particularly business-oriented scenarios, and connecting climate-related scenarios to business requirements; difficulties in quantifying climate-related risks and opportunities on business operations and finances; challenges around how to characterise resiliency. See TCFD (2019).

[13] *Impact scenarios* can also leverage existing *catastrophe models*. These models, developed in the last decades for insured losses, aim at identifying, assess and manage natural catastrophe risks linked in particular to seismic and climate hazards. Advanced models exist for tropical storms, floods, tornados or bushfires, mainly parametrised with historical data for specific countries or areas. These models can hence play a role in analysing physical risk for financial institutions, even though limitations remain. These latter are in particular due to the restrict number of countries and areas covered by existing models and the absence of structured relation between physical risk and financial risks.

it becomes possible to draw general estimates on the long-term value of the company, and the analysis based on these values can be expanded to create an aggregated assessment.

4.2.4   Setting the Risk-Taking Strategy

The management board is responsible for deciding the risk-taking strategy of the organisation (in line with the mandate given by the shareholders) and for allocating responsibility for its execution. To this extent, sustainability has to be considered as an emerging area of responsibility, for which the management board is today accountable *vis-à-vis* both shareholders and other stakeholders. Existing literature has already highlighted how, in dealing with sustainability issues, sustainability risk management is progressively emerging as a concrete business strategy option for organisations, which tries to align profit goals with internal climate, environmental and social policies (e.g. Anderson and Anderson 2009). Such policies seek to decrease the negative impact of the company by inter alia reducing the use of natural resources and decreasing carbon emissions, toxic substances (and by-products), and make this alignment efficient enough to sustain and grow the business while still preserving the environment and having a positive impact on the society (e.g. Anderson and Anderson 2009). In this respect, evidence also exists today proving that companies may derive tangible benefits (such as better client loyalty or the possibility to enforce a market premium for their products or services) from being perceived as engaged in sustainable activities (e.g. Anselmsson et al. 2014; Migliorelli and Dessertine 2019a), and that a generally positive relationship has existed at least in the last twenty years between corporate finance performances and environmental, social and governance (ESG) performances (e.g. Friede et al. 2015).[14]

However, new emerging elements, that literature has started exploring only recently, should be today considered by management boards when it comes to the impact of sustainability-related risks on financial risks. The first is the mismatch typically existing between the time horizon of the financial institution's risk-taking strategy and the time horizon of the sustainability-related risks (e.g. BoE 2015; BIS 2020). On the

---

[14] In many studies, this relationship has been proven even after taking into account the typical endogeneity problem (that is, the fact that the most financially successful companies may be the ones that decide to be involved in sustainable initiatives).

one hand, pressure from stakeholders on management boards for short-term profitability is often significative, consequently affecting strategic plans normally extending from three to five years only. On the other hand, sustainability-related risks are expected to progressively deploy their effects in the long-term (e.g. IPCC 2018) with, likely, a still relatively limited impact in terms of materiality in the short-term. As a result, management boards could tend to underestimate or have little interest in the potential long-term consequences of the sustainability-related risks implicitly accepted while adopting a specific risk-taking strategy.[15] Yet, some risk decisions typically result in a significant long-term engagement (e.g. in the case of the origination of a portfolio of 30 years mortgages or of a large investment launched to reach a new segment of clientele), and may reduce the flexibility of the organisation to react to changing market conditions. In such cases, locked-in effects may materialise when sustainability-related risks eventually start becoming material, with short-term payback possibly switching to long-term losses (see also Fig. 4.3).

A second element of attention for management boards when considering the impact of sustainability-related risks on financial risks is the fair reassessment of the financial institution's *risk appetite*. In point of fact, the voluntary exposure to a certain level of sustainability-related risks is in the strategic options available for the management board. In financial terms, this may mean for instance to accept higher credit risk via the exposure to transition risk (e.g. the case of banks deciding to keep providing lending to polluting sectors) or to accept higher market risk via the exposure to the reputational risk of major clients (e.g. the case of insurance companies having in their investment portfolios shares of companies not guaranteeing the fair treatment of workers). In this respect, it seems clear that the possible impacts of sustainability-related risks should become a structural part of the strategic reflection at the basis of the definition of the financial institution's *risk appetite*. To this extent, it can be argued that the organisation's capacity to systematically and accurately

---

[15] In a survey deliverd by Task Force on Climate-related Financial Disclosures (TCFD) to 198 financial institutions and other organisations, 60% of respondents said that their organisations consider climate-related issues to be a material risk today or in the next one–two years, while 28% consider climate-related issues to be a material risk only in six years or more, or were not sure about it. See TCFD (2019).

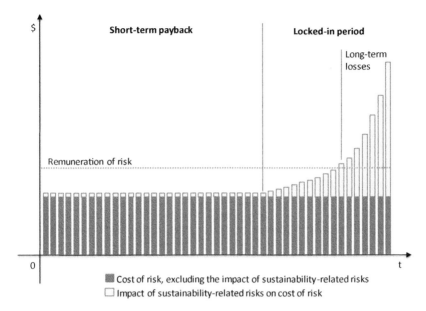

Fig. 4.3 Long-term impact of sustainability-related risks on short-seeing risk-taking strategy (*Notes* The figure shows a situation in which sustainability-related risks are not taken into account when deciding a short-seeing strategy. The cost of risk is not systematically adjusted for the long-term impact of sustainability-related risks. In such a situation, the increase in the materiality of the sustainability-related risks in the long-term [as part of the cost of risk], can result in short-term payback and long-term losses. *Source* Authors' elaboration)

price the sustainability-related risks will progressively increase in importance. In fact, the failure to correctly defining a pricing for these risks may result either in a non-remunerated risk-taking or in a missed opportunity following the choice of not accepting risks which cannot be duly measured and hence managed.

4.2.5 *Implementing the Risk-Taking Strategy*

Deal-level analysis is another key area of attention in the management of sustainability-related risks. In principle, when emerging risks are identified, they need to be reflected in the risk ratings of actual and potential clients. Nevertheless, little advances in this direction can be observed thus far in the financial industry as concerns sustainability-related risks (e.g. ACPR 2019; Oliver Wyman 2019). Several banks and other financial institutions have adopted specific policies for each of their sectors of activity in order to limit operations implying the financing of projects or businesses with negative climate, environmental or social impact (e.g. HSBC 2020).[16] As a matter of fact, this approach has been driven more by the financial institutions' new awareness of their critical role in fostering sustainability than by a concrete attention towards the management of the impact of sustainability-related risks on their financial risks (e.g. Migliorelli and Dessertine 2019b). In addition, such an approach does not properly shield financial institutions from the impacts of sustainability-related risks, as actual and potential clients may remain exposed to at least some of the sustainability-related factors (such as climate change or environmental degradation) largely independently from their sustainability performances.

Likely, a push towards the adjustment of the existing rating processes will come along with the consolidation of the materiality of sustainability-related risks. Most probably, this process will be first undertaken by established rating agencies and gradually embraced by financial institutions in their internal risk assessment activity (in particular by banks using internal rating systems). In the euro area, the European Securities and Markets Authority (ESMA) recently provided technical advice for the integration of sustainability into credit ratings (ESMA 2019a, b). In this respect, it can be noted that, in the matter of the integration of sustainability aspects in credit ratings, rating agencies are already used to consider governance-related aspects,[17] as they more directly influence strategic decisions, labour force policies and how business expansion is conceived (including in terms of polluting emission, conflicts of interests management, compliance, accounting and reporting practices, executives' compensation). Moreover, governance risks belong to every entity, whereas climate or environmental factors vary depending on the issuer's

[16] See here https://www.hsbc.com/our-approach/risk-and-responsibility/sustainability-risk the full set of sustainability risk policies adopted by HSBC, as a concrete example of exclusion criteria for non-sustainable projects or initiatives.

[17] At the same time, some credit rating agencies also include the regulatory environment, the legal and financial infrastructure, and the institutional environment in which governance takes decisions (ESMA 2019b).

industry sector and its location.[18] Concretely, approaches that can be adopted for assessing probability of default or expected losses refer to how sustainability-related factors affect the ability of the analysed entity to convert assets into cash, the impact of sustainability-related events on the entity's cost of capital, the impact of sustainability-management costs to the entity's profit generation (scenario analysis can again be used to this end). In proceeding in this direction of development, the main enablers, to be also pushed at policy level, are data availability and the consistency of both measurement and disclosure systems (ESMA 2019b).

4.3   DISCLOSURE OF SUSTAINABLE ACTIVITIES AND RELATED RISKS

4.3.1   *Materiality of Sustainability-Related Risks and Disclosure*

Financial disclosure aims at providing sufficient, reliable and comparable information on the company's performances to investors, and more in general to all types of stakeholders (including public administrations, citizens and non-profit organisations). In this way, an external unbiased assessment of the company becomes possible and investment decisions can be driven efficiently. In recent years, the disclosure of sustainability-related information (as part of non-financial disclosure) has started to be considered with growing interest by policymakers as an instrument potentially able to endogenously push private businesses to promote sustainable activities and policies.[19] Beyond its implicit cost, disclosing sustainability-related information can bring concrete benefits for companies. Among these benefits, it can be listed an improved corporate social responsibility awareness in the reference markets, an impulsion towards a better understanding of sustainability-related risks (including the aspect of risk management), and likely a more diverse investor base with a lower cost of capital.[20]

---

[18] When considered, environmental risks are sometimes proxied by renewable resources' usage and waste management, social risks are sometimes proxied by GDP per capita, income inequality, political risk, institutional strength and other indexes provided by accountable institutions such as the World Bank (ESMA 2019b).

[19] For an overview of the sustainability and environmental reporting evolution over time, see for example Weber and ElAlfy (2019).

[20] This can be a long-term result, for example linked to improved credit ratings for debt issuance and better credit worthiness assessments for bank loans, or to a better placement

When it comes to the identification of the sustainability-related information to disclose, even in a context featured by a high level of heterogeneity between jurisdictions and practices, some principles have progressively emerged. In this respect, disclosed information should be: material; fair, balanced and understandable; comprehensive but concise; strategic and forward-looking; stakeholder-oriented; consistent and coherent (e.g. EC 2019). As concerns the central aspect materiality, it can be easily argued that its assessment represents already a cornerstone in financial accounting and reporting, so that any item having financial impact *must* be measured and properly accounted for. Nevertheless, materiality disclosure related to sustainability factors should be considered today as having a (very) large scope, including not only information having direct and measurable impacts in financial terms, but also any element potentially influencing the value of the company in a broad perspective. By adopting this view, two major categories of information object of disclosure can be identified. The first refers to the activities carried out by the company and having either a positive or a negative sustainability impact. Examples are the disclosure of the level of greenhouse gas (GHG) emissions of the company or of the policies adopted on matters such as the fair treatment of employees, the respect of human rights, the anti-corruption and anti-bribery internal rules. The second category of information concerns the sustainability-related factors and risks that have, or are likely to have, a direct and measurable financial impact on the company.[21] Examples of this information are the quantification of the actual and potential losses due to the negative effects of climate change (e.g. in the form of high incidence of draughts, floods and storms) and the related list of locations or business lines that are subject to such risks.

The disclosure of sustainability-related information should hence allow a throughout understanding of the overall sustainability performances of the company and of the specific impacts of the sustainability-related risks. Such broad perspective is today relevant in particular for market investors, which indeed need to know about the sustainability characteristics of the companies for both portfolio selection needs (e.g. in the

---

of company securities in the portfolios of actively managed investment funds. However, such a possible outcome can count thus far on little evidence in scientific literature.

[21] The large part of the considerations included in this book refers to this second category.

case of sustainable funds picking shares of environmentally friendly businesses) and for financial risk assessments (e.g. in case of banks analysing the creditworthiness of clients under the threat of climate change).

4.3.2 Main Existing Standards and Frameworks

As of today, the approaches towards the disclosure of sustainability-related information remain widely heterogeneous, also as a consequence of the relative novelty of the matter. However, in the last years a number of jurisdictions and organisations have started to develop and promote standards or frameworks, focusing in particular on the impact of climate change, the preservation of the environment, the social and governance issues. While in same cases the disclosure of (certain types of) sustainability-related information has become mandatory (e.g. in the case of the EU Non-Financial Reporting Directive, NFRD), several initiatives have seen the light headed by international organisations and resulting in guidelines to steer voluntary disclosure. Noteworthy examples of the latter are the ones developed by the Climate Disclosure Project (CDP), the Climate Disclosure Standards Board (CDSB), the Global Reporting Initiative (GRI), the Principles for Responsible Investment (PRI), the Sustainability Accounting Standards Board (SASB) the Task Force on Climate-related Financial Disclosures (TCFD). These initiatives have had indeed the merit to contribute to foster best practices and promote the discussion on applicable standards both at policy and industry levels. Table 4.1 summarises the main disclosure standards and framework currently in place.

Among the mentioned standards and frameworks, the recommendations and the recommended disclosure of the TCFD have recently emerged as a particularly important reference as concerns climate change. The recommendations apply specifically to organisations in the financial sector, including banks, insurance companies, asset managers and asset owners, and are expected to be used in compliancy with existing mandatory disclosure obligations (in the jurisdictions in which such obligations are in place).[22] The TCFD has a strict financial materiality perspective and

---

[22] If certain elements of the recommendations are incompatible with national disclosure requirements for financial filings, the TCFD encourages organisations to disclose those elements in other official company reports that are issued at least annually, widely distributed and available to investors and others, and subject to internal governance processes that are the same or substantially similar to those used for financial reporting (TCFD 2017).

Table 4.1 Main disclosure standards and frameworks for sustainability-related information

Jurisdiction/organisation	Standard or framework	Sustainability perimeter	Key disclosure features
Jurisdiction European Union (EU)	EU Directive 2014/95 regarding disclosure of non-financial and diversity information (non-financial reporting directive—NFRD) and following guidelines on how to disclose	Social and environmental issues	Mandatory disclosure for listed companies (financial and non-financial). Reporting on policies implemented in relation to environmental protection, social responsibility and treatment of employees, respect for human rights, anti-corruption and bribery, diversity on company boards, greenhouse gas (GHG) emissions, energy consumption, energy production. The guidelines on how to disclosure are not binding

Jurisdiction/organisation	Standard or framework	Sustainability perimeter	Key disclosure features
France	Energy Transition Law (2015), Article 173	Climate change	Mandatory disclosure for listed companies (financial and non-financial). Disclosure of risks related to climate change, of the consequences of climate change on the company's activities and of the use of goods and services produced. For institutional investors, GHG emissions and contribution to the goal of limiting global warming
United States	Securities and Exchange Commission (SEC) guidance regarding disclosure related to climate change	Climate change	Mandatory disclosure for listed companies (financial and non-financial). Reporting on climate-related material risks and factors that can affect or have affected the company's financial condition, such as regulations, treaties and agreements, business trends and physical impacts

(continued)

Table 4.1 (continued)

Jurisdiction/organisation	Standard or framework	Sustainability perimeter	Key disclosure features
Organisation Carbon Disclosure Project (CDP)	CDP questionnaires	Climate change, forests, water security	Voluntary disclosure. Reporting information on governance, risks and opportunities, strategy, performance related to climate change, forests and water supply. Assessment based on scoring can also be included
Climate Disclosure Standards Board (CDSB)	Climate change reporting framework	Climate change	Voluntary disclosure. Reporting on the extent to which performance is affected by climate-related risks and opportunities, on the governance processes for addressing those effects, on the exposure to significant climate-related issues, on the strategy or plan to address the issues, on the level of GHG emissions

Jurisdiction/organisation	Standard or framework	Sustainability perimeter	Key disclosure features
Climate Disclosure Standards Board (CDSB)	Framework for reporting environmental information, natural capital and associated business impacts	Environment preservation	Voluntary disclosure. Reporting of environmental information that is material to understanding companies' financial risks. Information may include: natural capital dependencies, environmental results, environmental risks and opportunities, environmental policies and their outcome, performance against environmental targets
Global Reporting Initiative (GRI)	Sustainability reporting standards	Economic, environmental and social issues	Voluntary disclosure. Reporting standards for the organisation's significant economic, environmental and social impacts. Assessment include subjects related to materials, energy, water, biodiversity, emissions, waste, environmental compliance, suppliers
Principles for Responsible Investment (PRI)	PRI reporting framework	Economic, social and governance (ESG) issues	Voluntary disclosure. Reporting on the application of the principles for responsible investments

(continued)

Table 4.1 (continued)

Jurisdiction/organisation	Standard or framework	Sustainability perimeter	Key disclosure features
Sustainability Accounting Standards Board (SASB)	SASB conceptual framework, SASB standards	Environment, social capital, human capital, business model and innovation, leadership and governance	Voluntary disclosure. Disclosure materiality-driven, on issues that are likely to affect the financial condition or operating performance of companies within an industry. Standards developed for 77 industries
Task Force on Climate-related Financial Disclosures (TCFD)	TCFD recommendations and recommended disclosure	Climate change	Voluntary disclosure. Reporting on information related to governance, strategy, risk management practices, as well as metrics and target in relation to climate change-related issues (see also Table 4.2)

Source Authors' elaboration

divides climate-related risks into two major types: risks related to the transition to a low-carbon economy and risks related to the physical impacts of climate change (TCFD 2017). In this respect, recommendations and recommended disclosure are referred to the four key areas of governance, strategy, risk management, metrics and targets. One of the TCFD's key recommended disclosures refers to the resilience of an organisation when taking into consideration different climate-related scenarios, in particular a 2 °C (or lower) scenario. In this respect, the TCFD recognises the relevance of the use of scenarios in assessing climate-related issues and their potential financial implications, while also acknowledging that these practices are relatively recent and will progressively evolve over time (TCFD 2017). Table 4.2 details the TCFD recommendations and recommended disclosure.

4.3.3 Market Discipline and Sustainability-Related Risks: A Meaningful Relationship?

The question of whether sustainability-related disclosure is today an instrument able to effectively foster market discipline and efficiently drive investors decision is of the utmost importance. In this respect, even though corporate disclosure of sustainability-related information (for both financial and non-financial companies) has improved in recent years, it can be argued that significant gaps are still evident, and further improvements in the quantity, quality and comparability of disclosures is required to meet the needs of investors and other stakeholders. In particular, it can be observed that, even when mandatory disclosure requirements are foreseen, a high level of flexibility is usually granted to companies as concerns the type of information to disclose and how to disclose them. Such an approach, if on the one hand reduces the burden for companies in a necessarily introductory phase, on the other hand makes the comparison between companies often particularly difficult or even impossible. In point of fact, the high level of heterogeneity that can be today observed may dilute the effects of market discipline and hamper the widespread adoption of virtuous sustainability-oriented behaviours. In such a context, the location of the disclosure may also play a significative role. At best, the default location for the non-financial disclosure should be the company's management report or eventually the notes to financial statements. Nevertheless, many examples exist of companies taking the option (allowed by regulators) to publish their non-financial statement

Table 4.2   TCFD's recommendations and supporting recommended disclosures

Governance	Strategy	Risk management	Metrics and targets
Recommendations			
Disclose the organisation's governance around climate-related risks and opportunities	Disclose the actual and potential impacts of climate-related risks and opportunities on the organisation's businesses, strategy and financial planning where such information is material	Disclose how the organisation identifies, assesses and manages climate-related risks	Disclose the metrics and targets used to assess and manage relevant climate-related risks and opportunities where such information is material
Recommended disclosure			
• Describe the board's oversight of climate-related risks and opportunities • Describe management's role in assessing and managing climate-related risks and opportunities	• Describe the climate-related risks and opportunities the organisation has identified over the short, medium, and long-term • Describe the impact of climate-related risks and opportunities on the organisation's businesses, strategy, and financial planning • Describe the resilience of the organisation's strategy, taking into consideration different climate-related scenarios, including a 2 °C or lower scenario	• Describe the organisation's processes for identifying and assessing climate-related risks • Describe the organisation's processes for managing climate-related risks • Describe how processes for identifying, assessing, and managing climate-related risks are integrated into the organisation's overall risk management	• Disclose the metrics used by the organisation to assess climate-related risks and opportunities in line with its strategy and risk management process • Disclose on greenhouse gas (GHG) emissions, and the related risks • Describe the targets used by the organisation to manage climate-related risks and opportunities and performance against targets

Source Authors' elaboration on TCFD (2017)

in separate reports, also in the case of sustainability-related disclosure. Again, this approach can disperse the effectiveness of the disclosure. In this respect, it should be argued that when companies make use of separate reports they should at least ensure that the information is easily accessible for intended end-users, with the burden for the research and the usability of the information to be reduced at a minimum.[23]

References

Anderson, D., & Anderson, K. E. (2009). Sustainability risk management. *Risk Management and Insurance Review, 12*(1), 25–38.

Anselmsson, J., Bondesson, N. V., & Johansson, U. (2014). Brand image and customers' willingness to pay a price premium for food brands. *Journal of Product and Brand Management, 23*(2), 90–102.

Autorité de contrôle prudentiel et de résolution (ACPR). (2019). *French banking groups facing climate change-related risks.* Paris.

Bessis J. (2011). *Risk management in banking.* Chichester: Wiley.

Bank for International Settlements (BIS). (2020). *The green swan. Central banking and financial stability in the age of climate change.* Basel.

Bank of England (BoE). (2015, September). *Breaking the tragedy of the horizon—Climate change and stability.* Speech given by Mark Carney, Governor of the Bank of England.

Banque de France. (2020). *Long-term growth impact of climate change and policies: The Advanced Climate Change Long-term (ACCL) scenario building model* (Working Paper Series no. 759).

Christoffersen, P. F. (2003). *Elements of financial risk management.* Amsterdam: Academic Press.

De Servigny, A., & Renault O. (2004). *Measuring and managing credit risk.* Ne York: McGraw-Hill.

European Commission (EC). (2018). *A Clean Planet for all. A European strategic long-term vision for a prosperous, modern, competitive and climate neutral economy.* COM(2018) 773 final, Brussels.

European Commission (EC). (2019). *Guidelines on non-financial reporting: Supplement on reporting climate-related information.* 2019/C 209/01, Brussels.

---

[23] To this extent, the TCFD proposes that its recommended disclosures should be included in the company's mainstream *annual financial filings.* If companies make cross-references to other reports or documents, this should be done in a simple and user-friendly way, for instance, by applying a practical rule of "*maximum one click*" out of the report.

European Securities and Markets Authority (ESMA). (2019a). *Final report—Guidelines on disclosure requirements applicable to credit ratings*. ESMA 33-9-320.

European Securities and Markets Authority (ESMA). (2019b). *ESMA Technical advice to the European Commission on sustainability considerations in the credit rating market*. ESMA 33-9-321.

Friede, G., Busch, T., & Bassen, A. (2015). ESG and financial performance: Aggregated evidence from more than 2000 empirical studies. *Journal of Sustainable Finance and Investment, 5*(4), 210–233.

HSBC. (2020). *Introduction to HSBC's sustainability risk policies*. London.

Intergovernmental Panel on Climate Change (IPCC). (2018). *Special report. Global warning of 1.5 °C*.

International Energy Agency (IEA). (2019). *World Energy Model. Scenario of future energy trend*. Paris.

Jorion, P. (2007). *Value-at-risk*. New York: McGraw-Hill.

Migliorelli, M., & Dessertine, D. (2019a). From transaction-based to mainstream green finance. In M. Migliorelli & D. Dessertine (Eds.), *The rise of green finance in Europe. Opportunities and challenges for issuers, investors and marketplaces*. Cham: Palgrave Macmillan.

Migliorelli, M., & Dessertine, D. (Eds.). (2019b). *The rise of green finance in Europe. Opportunities and challenges for issuers, investors and marketplaces*. Cham: Palgrave Macmillan.

Oliver Wyman (2019). *Climate change. Managing a new financial risk*. London.

Potsdam Institute for Climate Impact Research (PIK). (2014). *Understanding change in patterns on vulnerability* (PIK Report 127).

Prudential Regulation Authority (PRA). (2018, September). *Transition in thinking: The impact of climate change on the UK banking sector*.

Task Force on Climate-related Financial Disclosures (TCFD). (2017). *Recommendations of the Task Force on Climate-related Financial Disclosures*. Basel.

Task Force on Climate-related Financial Disclosures (TCFD). (2019). *Status report*. Basel.

Weber, O., & ElAlfy, A. (2019). The development of green finance by sector. In M. Migliorelli & D. Dessertine (Eds.), *The rise of green finance in Europe. Opportunities and challenges for issuers, investors and marketplaces*. Cham: Palgrave Macmillan.

CHAPTER 5

Sustainability-Related Risks and Financial Stability: A Systemic View and Preliminary Conclusions

Marco Migliorelli, Nicola Ciampoli, and Philippe Dessertine

Abstract This chapter discusses the possible impact of sustainability-related factors (such as climate change, environmental degradation, social inequality, policy and technology shifts) on financial stability. To this extent, it first identifies the areas in which an evolution of the practices

The contents included in this chapter do not necessarily reflect the official opinion of the European Commission. Responsibility for the information and views expressed lies entirely with the authors.

M. Migliorelli (✉) · P. Dessertine
IAE Université Paris 1 Panthéon-Sorbonne (Sorbonne Business School), Paris, France
e-mail: Marco.Migliorelli@ec.europa.eu

P. Dessertine
e-mail: dessertine.iae@univ-paris1.fr

M. Migliorelli
European Commission, Brussels, Belgium

© The Author(s) 2020   119
M. Migliorelli and P. Dessertine (eds.), *Sustainability and Financial Risks*, Palgrave Studies in Impact Finance,
https://doi.org/10.1007/978-3-030-54530-7_5

of financial intermediaries are necessary to better manage sustainability-related risks. This refers in particular to the existing risk-management frameworks (which may not consider sustainability-related risks) and to the timespan of the risk-taking strategies (which typically underestimate the long-term nature of sustainability-related risks). Hence, the chapter discusses a set of policy actions to both mitigate and control for sustainability-related risks. In this respect, it focuses on the need of evolving the prudential supervisory approaches and on the possibility to assign a more active role to central banks.

Keywords Sustainability-related risks · Climate change-related risks · Financial stability · Risk management frameworks · Macroprudential supervision · Central banks

5.1   Introduction

Financial stability can be defined as the condition in which the financial system, comprising financial intermediaries, markets and market infrastructures, is capable of withstanding shocks and the unravelling of financial imbalances (ECB 2019). In such a condition, the likelihood of disruptions in the financial intermediation process that are systemic and severe enough to trigger a material contraction of real economic activity can be considered to be mitigated. Already before the 2007–2009 Great Crisis, literature had highlighted three key characteristics that a financial system needs to have in order to maintain financial stability (e.g. Fell and Schinasi 2005). First, it has the capacity to efficiently and smoothly facilitate the intertemporal allocation of financial resources from savers to investors. Second, it can comfortably absorb both financial and real economic shocks. Third, it foresees mechanisms and practices to ensure that financial risks are assessed, priced and managed accurately by financial intermediaries. If one or more of these characteristics is not present, then it is likely that the financial system is moving in the direction of

N. Ciampoli
LUMSA University, Rome, Italy
e-mail: n.ciampoli@lumsa.it

becoming less stable, and at some point in time it might exhibit instability. Monitoring and ensuring financial stability implies the throughout considerations of a full range of potentially harming factors, both external and internal to the financial intermediaries, including an assessment of the systemic relevance of the potential fragilities coming from each financial actor (this latter in particular to avoid contagion effects[1]).

When it comes to the possible impact of sustainability-related risks on financial stability, it is relevant to analyse the matter under the light of the possible emergence over time of new (and underestimated) financial risks. This chapter mainly focuses on this issue. To do that, it is structured as follows. First, it introduces the key features of the policy frameworks typically in place to ensure financial stability, as well the possible role of sustainability-related risks in these frameworks. This is done in Sects. 5.2 and 5.3. Second, it proposes a number of actions to be undertaken by policymakers to effectively shield financial stability from sustainability-related risks, with particular emphasis on the role of prudential supervisory authorities. This analysis is carried out in Sects. 5.4, 5.5 and 5.6. Finally, some preliminary conclusions on the relationship between sustainability-related risks and financial stability are presented in Sect. 5.7.

5.2 Brief Outline of the Policy Approach to Financial Stability

Given its broad scope, financial stability has been traditionally pursued by policymakers worldwide through a structured mix of regulation and organisational structures. To this extent, a preventive arm of policy action aiming at limiting situation of crisis is normally accompanied by a remedial harm dealing with specific cases of financial distress. The preventive arm typically includes the definition of prudential legislation, the empowerment of authorities and agencies with specific supervisory and regulatory

---

[1] In this respect, the Basel Committee on Banking Supervision (BCBS) published in October 2012 a principles-based framework for dealing with domestic systemically important banks (BCBS 2012). The European Union implemented this framework in the Capital Requirements Directive (CRD IV) and the European Banking Authority adopted guidelines that recommend to the national macroprudential authorities the approach to follow for the identification of systemically important banks at the domestic level. Hence, national authorities designate systemically important banks and set capital buffers for them.

competences, the public support to the establishment of a sound physical market infrastructure (and related market conventions), as well as the periodical provision of information by public authorities to the market on the most relevant economic and financial risks. On the other hand, the remedial arm usually foresees different forms of liquidity and solvency support for distressed financial entities as well as mechanisms to orderly managing deeper crises, including in case of entities' resolution.[2,3]

Prudential supervision represents in such a context a central pillar for the safeguard of financial stability, and it is particularly relevant when it comes to the analysis of the systemic impacts of sustainability-related risks on financial risks. In broad terms, prudential supervision refers to the oversight of financial institutions to ensure that they are complying with relevant regulation and, more generally, are operating soundly and prudently in line with the principles laid down by the financial stability framework.

In the last two decades, it has been observed a tendency to assign prudential supervision of credit and insurance institutions to central

[2] See for example Allen and Wood (2006).

[3] An example of the structured mix of the regulation and organisational structure in place to ensure financial stability is the one adopted in the European Union (EU). In the EU, financial stability is first nested in the framework defined by the combination of the so-called Banking Union and Capital Markets Union. The Banking Union is mainly built around the Capital Requirements Directive and Capital Requirements Regulation (CRD IV-CRR package, derived by the Basel Accords), the Bank Recovery and Resolution Directive (BRRD) and the Deposit Guarantee Schemes Directive (DGSD). The overall aim of these pieces of legislation is to enforce financial stability through a mix of measures designed to both reduce and share banking sector risks. In this respect, the Banking Union eventually results to be based on three pillars: a single supervisory mechanism (SSM), a single resolution mechanism (SRM) with a related single resolution fund, and a European deposit insurance scheme (EDIS). However, the EU macro-prudential framework is to a significant extent implemented in a decentralised way. Authorities in the Member States identify risks and may implement macro-prudential measures within the remit of their jurisdiction. Such a decentralised set-up is mainly due to the fact that systemic risks are often local or national in scope and interrelate with specific national situations (see for example EC 2019a). To balance this decentralised implementation, the EU macroprudential framework also comprises mechanisms to avoid excessive heterogeneity. To this extent, the European system of financial supervision (ESFS) was introduced in 2010. It

banks, as a supplementary activity to the one of definition and implementation of the monetary policy.[4] A set of instruments is usually in the toolkit of prudential supervisory bodies to perform their tasks, with different instruments used for different purposes. As concerns the supervision of credit institutions, specific instruments are aimed at facing the cyclical systemic risks which may arise from the self-perpetuating interactions between lending, on the one hand, and the valuation of the real and financial assets used as collateral, on the other hand (the relationship between mortgages lending practices and the price of real estate is a typical example). Scenario analyses and stress testing are widely used techniques to monitor the potential magnitude of these types of risks. Other instruments look at the broad structure of the financial markets and are intended to attenuate the risks arising from the dominant positions that some institutions may acquire or can result from a high level of interdependence between financial institutions. The request of supplementary capital buffers for systemic relevant banks is a case of possible risk mitigating action in such cases (e.g. NBB 2018). Finally, other instruments exist dealing with specific risks. These are used in particular in the management of liquidity and capital positions in banking groups, or in the adjustment of the capital requirements in line with specific developments in the financial or property markets (e.g. NBB 2018).

comprises the European Systemic Risk Board (ESRB), which ensures that the objectives of financial integration at EU level and financial stability at the Member State level can be jointly pursued, and the three European supervisory authorities (ESAs), namely: the European Banking Authority (EBA), the European Securities and Markets Authority (ESMA) and the European Insurance and Occupational Pensions Authority (EIOPA). On the other hand, financial stability is also fostered through the implementation of the Capital Markets Union, the blueprint headed by the European Commission to channel financial resources to all types of companies and infrastructure projects that need it to expand and create jobs. A first specific action plan to build the Capital Market Union has been published in 2015 (see EC 2015).

[4] Central banks act to pursue specific objectives as defined by their statutory documents. For example, the main objective of the European Central Banks (ECB) is to maintain price stability, defined as a year-on-year increase in the Harmonized Index of Consumer Prices (HICP) for the euro area close but below 2%. In addition, without prejudice to the objective of price stability, the ECB may support the general economic policies in the Union. These may include, inter alia, full employment and balanced economic growth.

5.3   FINANCIAL INTERMEDIARIES AND KEY
AREAS OF ATTENTION IN THE MANAGEMENT
OF SUSTAINABILITY-RELATED RISKS

In point of fact, an argument can be made according to the idea that the threat to financial stability coming directly from sustainability-related risks is somewhat limited in the short-term. The likelihood of the appearance of a systemic crisis in the financial sector in the next few years stemming from factors such as climate change, environmental degradation, social inequality or policy and technology shifts is reasonably low.[5] However, such a threat is expected to considerably increase in the future with the foreseen consolidation of sustainability-related risks, both in terms of frequency and magnitude (the severe consequences of the expected increase in temperature worldwide in the next decades are an example of this dynamic[6]). For this reason, the implementation of forward-looking mitigation actions by both supervisory entities and financial institutions in order to preserve financial stability from sustainability-related risks should not be postponed too late in the future.

In this respect, at least two (interrelated) elements featuring the approach to risk-management and risk-taking of financial intermediaries should be carefully considered in the assessment of the possible impact of sustainability-related risks on financial stability. On the one side, the reliability of the existing risk-management frameworks, which may not systematically and coherently take into account the occurrence of sustainability-related risks and their link to financial risks. On the other side, the possible timing mismatch between the risk-taking strategies of financial intermediaries, often shaped to produce results in the short or medium term, and the profile over time of the sustainability-related risks, these latter expected to have increasingly disruptive effects in the longer term (e.g. BoE 2015; ACPR 2019; BIS 2020).

The need for financial intermediaries to adjust their risk management frameworks in order to consider the possible impact of sustainability-related risks on their financial risks may progressively become material. This implies the incremental development of specific risk management

---

[5] For a wider analysis of the relation between sustainability-related risks and financial risks, see Chapter 1.

[6] See IPCC (2018) or BIS (2020).

methodologies[7] and, more in general, an enrichment of the risk culture. In particular, models able to predict the development of climate and environmental variables and simulate economic impact scenarios tailored on the business of each intermediary (specifically in terms of geographic areas and economic sectors served) would need to be progressively developed. In addition, as risk-taking activities are typically decentralised at operational units, the familiarity of all the relevant levels of the organisation with the possible consequences for the business of sustainability-related factors is also expected to play an increasingly important role.

On the other hand, it would be necessary for financial actors to critically reassess the reference horizon of their risk-taking strategies, in a way to be able to fully factoring in sustainability-related risks (e.g. BoE 2015). As a matter of fact, the tendency of part of the financial industry to privilege short-termism and strategies set to produce positive returns in the space of a few years clearly goes in the opposite direction of a throughout understanding of sustainability-related risks, which in many cases spread their effects in the long-term.[8] In this respect, it is important to underline that threats to financial stability may also arise from the possibility of disorderly adjustments of imbalances that have built up endogenously over a period of time because, for example, expectations of future returns were misperceived and therefore mispriced. This latter can be in particular the case for stranded assets and as a consequence of policy and technology changes.

5.4 THE WIDER POLICY APPROACH

In general terms, two major areas of action can indeed be highlighted for policymakers in order to shield financial stability from the expected impact of sustainability-related risks. The first concerns the set of actions to put in place with the aim of limiting the foreseen incidence of these risks. The second refers to the most effective ways to effectively dealing with the effects of the risks.

The possibility of limiting the expected incidence of sustainability-related risks on financial risks does not lay only in the field of finance. On the contrary, it implies the prior assessment of the different facets of the sustainability of human activities on the planet as well as the conception

---

[7] For further details, see Chapter 4.

[8] See, as concerns global warming and climate change, IPCC (2018).

and evaluation of the possible corrective measures. As a matter of fact, the possibility to trigger a trend reversal in matters such as climate change or environmental degradation depends on a multitude of factors. It can be expected that aspects regarding political engagement, effective regulation, technological improvements, scientific research and investment flows will jointly determine the feasibility and speed of the changeover (e.g. EC 2019b; Migliorelli and Dessertine 2019). For this reason, actions aiming at reducing the sustainability-related risks should indeed be nested in the wider policy initiatives having as objective the fight against climate change, the restoration and preservation of the environment, the reduction of the inequalities. On the other hand, it can be argued that policy actions aimed at specifically dealing with the financial risks stemming from sustainability-related risks (that is understanding, measuring and managing the relationship between the different types of sustainability-related risks and financial risks) can be more easily narrowed in scope, and specifically assessed in the traditional perimeter of action of prudential supervision authorities.

In such a context, Table 5.1 reports six key policy actions that jointly would likely allow to effectively limit and deal with the sustainability-related risks and, in turn, would also safeguard financial stability. In point

Table 5.1   Key policy actions to safeguard financial stability from sustainability-related risks

#	Action	Objective
1	Implementing the Paris Agreement and reaching the Sustainable Development Goals	Reducing the risk
2	Mainstreaming sustainable and green finance	Reducing the risk/dealing with the risk
3	Assessing the impact of climate policies in order to limit sideeffects	Reducing the risk
4	Fostering economic resilience to sustainability-related risks	Reducing the risk
5	Ensuring prudential supervision of the impact of all sustainability-related risks on financial markets	Dealing with the risk
6	Establishing global governance structures for the analysis of the impact of sustainability-related risks on financial markets	Dealing with the risk

Source Authors' elaboration

of fact, adopting such a set of actions would results first and foremost in the definition of a comprehensive policy programme backed by a strong commitment towards the achievement of a sustainable society.

5.4.1   Implementing the Paris Agreement and reaching the Sustainable Development Goals[9]

The Paris Agreement and the Sustainable Development Goals, both dated 2015, are today the cornerstones of the international community's engagement towards the fight against the climate change and the construction of a more sustainable and fair society. However, it is increasingly evident that the fortune of these deals will decisively depend on the level of political commitment during the (necessary long) implementation phase.[10] Some jurisdictions, such as the European Union, have already made some concrete steps to drive the change,[11] while in other cases, as for the United States, a certain level of disengagement has been observed.[12] Intermittent political commitment can risk to dilute the efforts and the results reached thus far. On the contrary, a full implementation of the Paris Agreement and the achievement of the Sustainable Development Goals will allow a sensitive reduction of all the sustainability-related risks, in this way also reducing their potential impact on financial stability.

[9] For a wider dissertation on the Paris Agreement and the Sustainable Development Goals, see Chapter 1 or Berrou et al. (2019b).

[10] In particular for the Paris Agreement, the political commitment of the countries responsible for the largest part of the greenhouse gas emission is essential. In this respect, China counts for about 26% of the GHG emissions, the U.S. for 15%, the EU for 10%, India for 6%, Russia for 5%, Japan for 3%, Brazil for 2%. Source: World Research Institute. Data referred to 2014.

[11] At European Union's level, it should at least be listed the issuance of the European strategic long-term vision for a prosperous, modern, competitive and climate-neutral economy (EC 2018a) and the European Green Deal (EC 2019b).

[12] In June 2017, United States President Donald Trump announced his intention to withdraw his country from the Paris Agreement. Under the agreement itself, the earliest effective date of withdrawal for the United States is November 2020.

5.4.2   Mainstreaming Sustainable and Green Finance

Sustainable finance may be referred to the process of taking due account of environmental and social considerations in investment decision-making, leading to increased investments in longer term and sustainable activities (EC 2018b).[13] In this respect, green finance can be considered to be part of the wider sustainable finance landscape.[14] The growth of sustainable finance market, and in particular of green finance, can be considered today robust, as increasing volumes are accompanied by sectorial diversification and a continuously widening range of products. Nevertheless, it can be argued that the actual levels of issuance of sustainable securities is still nothing more than "a drop in the ocean" when it is compared to estimated needs for an effective financing of the sustainability objectives.[15] As a matter of fact, for sustainable finance to effectively contribute to mitigate the sustainability-related risks, from a promising niche it has to evolve in a mainstream way of financing. To do that, the full involvement of key policymakers is fundamental as market forces alone will most probably be ineffective to produce the necessary transition.[16]

5.4.3   Assessing the Impact of Climate Policies in Order to Limit Side Effects

Climate policies in particular are expected to trigger an unprecedented shift in the structure of the economies that will embrace the change. If effective, these policies will be also accompanied by a new stream of

---

[13] For more details on possible definitions of sustainable finance and green finance, see UNEP (2016) or Berrou et al. (2019a).

[14] See also Chapter 1.

[15] As an example, investments of around EUR520–575 billion annually have been estimated to be necessary in the EU in order to achieve a net-zero greenhouse gas economy in the 2050 horizon (EC 2018a). In 2018, annual emissions of labelled green bonds (the major security in the green finance market segment) in the EU can be estimated in less than EUR50 billion (authors' elaboration on data CBI).

[16] Namely, to mainstream sustainable finance it would be needed that environmental and other sustainability-related risks are properly included in the investors' decision-making processes, market demand is effectively channelled towards sustainable investments, additionality is adequately encouraged by policymakers when needed and the banking sector is fully engaged in the transition. For a more detail dissertation, see also Migliorelli and Dessertine (2019).

technology advances that will help reducing greenhouse gas emission and eventually minimise the dependency from fossil fuels. Policy and technology shifts may however produce considerable side effects as they can result in a rapid and unaffordable obsolescence of a large amount of economic assets.[17] A typical case could be represented by the loss of value of oil companies in a society with 100% of electric cars and green buildings. As a matter of fact, financial stability could also be threatened by policies aiming at fostering sustainability without considering the consequent impact on traditional sectors and disrupted incumbent industries. Hence, it is necessary for climate policies to be backed by an assessment of potentially negative impacts on specific economic actors, and foresee when necessary adapted mitigating and transition measures (e.g. in the form of "transition funds").

5.4.4 *Fostering Economic Resilience to Sustainability-Related Risks*

Financial stability may increasingly depend on the economic resilience to sustainability-related risks of the different economic actors. In this respect, the strengthening of the response capabilities to climate and environmental risks in particular (which can eventually translate into physical risks and other risks arising from natural catastrophes) should be considered a key policy objective in the years to come. In principle, such resilience should be endogenously built over time by economic actors, in particular by continuing assessing the potential impacts of sustainability-related risks on their businesses, and planning investments consequently. Nonetheless, the role of public actors in this area is still of the utmost relevance and should consist in at least three concrete actions. First, to raise awareness towards relevant stakeholders on the expected increasing incidence of sustainability-related risks on their business, that can be underestimated due to short-seeing approaches to risk-taking and focus on the specific phase of the business cycle. Second, to ensure that key national and international infrastructures are resilient to sustainability-related risks (and in particular to the ones linked to climate change), in order to avoid major disruptions in trade and business operations. Third, to foster market

---

[17] Estimates of losses are large and range from USD1 trillion to USD4 trillion when considering the energy sector alone (IAE and IRENA 2017).

discipline and allow better pricing and monitoring of the exposure of businesses to sustainability-related risks, in particular by identifying and set specific disclosure requirements for these risks.[18]

5.4.5   *Ensuring Prudential Supervision of the Impact of All Sustainability-Related Risks on Financial Markets*

As of today, only a few authorities in charge of financial stability have started to study the possible incidence of sustainability-related risks on financial risks, by focusing on climate change-relate risks and, at a lesser extent, to environmental-related risks (e.g. PRA 2018; ACPR 2019). As the potential impact of sustainability-related risks become more accepted and material, the need for a more structured approach under a prudential supervision perspective also materialises. Among the suitable actions, systematically monitoring the overall exposure and resilience of the financial system to sustainability-related risks and encouraging financial intermediaries to develop specific methodologies for handling such risks conveys particular importance. In addition, an assessment of the existing prudential requirements can also be needed. In fact, it can be argued that existing capital requirements may largely play against the full inclusion of sustainability-related risks in risk management frameworks and increase the possibility of market failures. As the Basel framework adopt a risk-weighted approach to capital consistencies, banks may need to bear increasing capital costs when fully considering sustainability-related risks. Hence, an effort to better integrate such risks into prudential supervision frameworks by verifying the suitability of the existing capital requirements may be also necessary in the mid-term.[19]

---

[18] See also Chapter 4.

[19] The capital requirements set out in the Pillar 1 of the Basel III framework do not consider sustainability-related risks directly (capital is explicitly required only for credit and operational risk s related to borrowers that violate environmental regulations), so it can be argued that the Basel III framework is not adapted as such to promote a progressive integration of sustainability-related risks (Cambridge and UNEPFI 2014). Despite the fact it seems attractive to foster green lending by regulatory-based incentives linked to Pillar 1 (e.g. by lowering risk weights or by using other types of "green supporting factors"), the prudential regime should remain fully focused on risk management and any innovation carefully assessed. Weak material incentives (e.g. slightly lowered risk weights for sustainable assets) would probably not change the banks' investment behaviour, whereas greater incentives may have the undesired effect to incentivise regulatory arbitrage towards exposures that absorb less regulatory capital while still bearing financial risk and existing

5.4.6 Establishing Global Governance Structures for the Analysis of the Impact of Sustainability-Related Risks on Financial Markets

The establishment of global structures for the fair assessment of the possible impacts of sustainability-related risks on financial markets can produce considerable benefits when it comes to the need to identify the most effective ways forward to deal with these risks. In point of fact, sustainability-related risks are often originated from the cumulative behavior of actors located in several different countries or even continents. In addition, the response to the threats coming from sustainability-related factors may not be effective if not implemented globally (the fight against the increase of temperatures due to global greenhouse gas emissions is a typical example). A global governance for the sustainability-related risks implies the creation of specific bodies or agencies empowered to discuss relevant items (such as regulation effectiveness, data gathering and sharing, methodological approaches, standards for disclosure) and formulate policy recommendations. The Network for Greening the Financial System (NGFS),[20] the Sustainability Accounting Standard Board (SASB)[21] and the Task Force on Climate-related Financial Disclosures (TCFD)[22] represent noteworthy initiatives in this direction, even if limited in scope.

regulatory uncertainties (e.g. related to the definition of sustainable, green or brown assets).

[20] Launched at the One Planet Summit in Paris in December 2017 under the initiative of the Banque de France, the NGFS is composed by more than 30 central banks, supervisory bodies and international organisations (including Banco de España, Bank of England, Bank of Finland, Banque Centrale du Luxembourg, Deutsche Bundesbank, European Banking Authority, European Central Bank, Japan FSA, National Bank of Belgium, Oesterreichische National Bank, the People's Bank of China, the Reserve Bank of Australia, Reserve Bank of New Zealand). It aims on a voluntary basis to exchange experiences and best practices, to contribute to the development of environment and climate risk management in the financial sector, and to mobilise mainstream finance to support the transition towards a sustainable economy. In 2019, the NGFS issued the first comprehensive report on climate change as source of financial risk (NGFS 2019).

[21] The SASB (https://www.sasb.org/) is a non-profit organisation that sets financial reporting standards on the issue of sustainability. In this respect SASB standards have as objective to enable businesses to identify, manage and communicate financially material sustainability information to their investors.

[22] The TCFD (https://www.fsb-tcfd.org/) aims at developing voluntary, consistent climate-related financial risk disclosures for use by companies in providing information to

5.5   SUSTAINABILITY-RELATED RISKS
AND PRUDENTIAL SUPERVISION

In the last a few years only, first progress has been made by financial stability authorities in understanding how the financial system may be vulnerable to the physical risk of climate change and to risks from a slow response to the need for a transition to an economy with lower carbon emissions (ECB 2019). However, these authorities, which include in particular central banks, still face significant gaps in the availability of assessments of risk management and stress testing frameworks, as well as in the availability of comprehensive and reliable disclosures and the reporting of relevant data, such as carbon emission-related data (ECB 2019; BIS 2020).

As mentioned above, prudential supervision practices should evolve in order to take into account the novelties introduced by sustainability-related risks.[23] In this respect, the NGFS provided in 2019 a high-level roadmap for the integration of climate-related factors into prudential supervision, highlighting a possible course of action. The actions suggested refer to raising awareness and building capacity among firms, assessing climate-related risks, setting supervisory expectations, requiring

---

investors, lenders, insurers and other stakeholders. The TCFD in particular considers the physical, liability and transition risks associated with climate change and what constitutes effective financial disclosures across industries.

[23] In April 2020 Basel Committee published the main results of a survey (BCBS 2020) on the initiatives on climate-related financial risks conduct on 27 Basel Committee members, including the European Central Bank (ECB) and the European Banking Authority (EBA). A large majority of these supervisors detected that they do not have an explicit mandate with regards to climate-related financial risks, but indicated that such risks could potentially impact the safety and soundness of individual financial institutions and could pose potential financial stability concerns for the financial system. Then, these institutions believe they can act within their existing mandate to mitigate climate-related financial risks. Even if the climate-related financial risks are not specifically designated in their regulatory and supervisory framework, most of these supervisory authorities consider these risks to fall implicitly within their existing framework, since the existing prudential framework requests banks to manage all risks of relevance, including climate-related financial risks. However, a few authorities are of the view that climate-related financial risks should be manifested or embedded into the existing risk categories (e.g. credit risk, operation risk, etc.), rather than be considered a new and standalone category of risk. Less than half of the Basel Committee members have issued dedicated supervisory guidance related to the governance, strategy and/or risk management of climate-related financial

transparency to promote market discipline, mitigating risk through financial resources. The full roadmap and related measures is reported in Table 5.2.[24]

Such a comprehensive approach has indeed the merit to fully consider climate-related risks as elements having an impact on financial risks, and to propose a way forward to gradually put in place a structured prudential supervisory framework to the management of these risks. However, the NGFS's recommendations are not compulsory and the reach of this body is indeed not global (in particular, the United States have not taken part to this initiative). Hence, a certain level of heterogeneity in the responses to this call for action should be expected. In addition, the limitation to climate-related factors (that is, not including among others environmental, social, policy and technology factors) still makes the full management of the impact of sustainability-related risks on financial risks an objective far to be reached. In this respect, the gradual extension of prudential supervisory action to other sustainability-related risks should be encouraged.

risks by banks. The form chosen to this supervisory guidance is guidelines, action plans or supervisory statements, and they are not always legally binding rules. Rather, it is principle-based or interpretations of existing rules. In addition to supervisory guidance, some institutions are working on identifying 'best practices' to mitigate climate-related financial risks and some of these initiatives are being conducted together with private-sector participants. Most supervisors have not yet included some form of the mitigation of climate-related financial risks into the prudential capital framework. However, some institutions are still quite far from being able to quantitatively assess the climate-related financial risks in the context of capital. As such, they have no short-term plans to consider applying Pillar 1 or Pillar 2 requirements for climate-related financial risks. Regarding the potential application of Pillar 2 capital add-ons, several institutions believe that the current Pillar 2 framework offers flexibility to address climate-related financial risks. Under Pillar 2, banks are required to develop the internal capital adequacy assessment process to capture all material risks that are not sufficiently covered under Pillar 1. Such risks would also include climate-related financial risks if they are estimated to be material to the specific financial institution.

[24] The NGFS also provided a set of six recommendations to enhance the role of central banks, supervisors, policymakers and financial institutions in the greening of the financial system and the managing of and climate change and environment-related risks (NGFS 2019). Namely: (i) integrating climate-related risks into financial stability monitoring and

Table 5.2   High-level roadmap for the integration of climate-related factors into prudential supervision

Course of action	Possible measures by supervisors
Raising awareness and building capacity among firms	• Raise awareness of the relevance of climate-related risks publicly and during bilateral meetings; survey firms on the impact of these risks; lay out a strategic roadmap for the handling of climate-related risks • Build capacity by convening events to progress the translation of scientific findings to financial analysis; set up working groups with firms, for example, on incorporating climate issues into risk management or scenario analysis
Assessing climate-related risks	• Develop analytical tools and methods for assessing physical and transition risks related to climate change both at a micro- (financial institutions) and macro-level (e.g. the financial system) • Conduct and publish an assessment of these risks at a macro- and micro-level • Analyse potential underlying risk differentials of "green" and "brown" assets. This pre-supposes that the supervisor and/or jurisdiction have agreed on definitions and classifications for "green" and "brown" activities
Setting supervisory expectations	• Issue guidance on the appropriate governance, strategy and risk management of climate-related risks by regulated firms • Train supervisors to assess firms' management of these risks
Requiring transparency to promote market discipline	• Set out expectations for firms' climate-related disclosures in line with the Task Force on Climate-related Financial Disclosure (TCFD) recommendations[a] • Consider integrating climate-related disclosure into Pillar 3 [of the Basel framework]

(continued)

Table 5.2 (continued)

Course of action	Possible measures by supervisors
Mitigating risk through financial resources	• Consider applying capital measures in Pillar 2 [of the Basel framework] for firms that do not meet supervisory expectations or with concentrated exposures • Based on the risk assessment outlined above, possibly consider integrating it into capital requirements of Pillar 1 [of the Basel framework]

[a]See TCFD (2017)
Source Authors' elaboration on NGFS (2019)

5.6   A New Role for Central Banks?

Central banks, in particular in Europe, are gradually emerging as critical actors in the policy action aiming at dealing with the potential financial risks coming from climate change (e.g. BoE 2015; PRA 2018; ACPR 2019), in this way paving the way for a better comprehension of the sustainability-financial risk nexus. Such dynamism has been linked to their prudential supervision mandate and it mainly results in assessing climate risks as a new source of financial risk potentially able to harm financial stability. In this vein, in the most ambitious approaches, the possibility to run "climate change stress tests" is under discussion in order to measure the resilience of financial intermediaries to different climate scenarios.

Nevertheless, when considering the possible role of central banks in ensuring that financial stability is not affected by sustainability-related risks, an additional dimension of action might be discussed. This refers to the possible extension of the central bank's mandate to formally include the support to the attainment of the sustainability objectives. Such an option presents considerable potential benefits, but also engenders some

micro-supervision; (ii) integrating sustainability factors into own-portfolio management; (iii) bridging the data gap; (iv) building awareness and intellectual capacity and encouraging technical assistance and knowledge sharing; (v) achieving robust and internationally consistent climate and environment-related disclosure; (vi) supporting the development of a taxonomy of (environmentally sustainable) economic activities. These recommendations are not binding and reflect the best practices identified by NGFS members.

concrete risks. In practice, this would probably mean to integrate sustainability considerations in the implementation of the monetary policy, by setting specific eligibility criteria for the securities object of the central bank's open market operations of assets purchase or for the banks' marginal deposit operations towards the central bank. Part of the central banks' operations (in terms of a fixed portion, a ceiling or a floor) might hence be reserved to sustainable securities. In this respect, the eligibility criteria to be fixed would eventually need to mirror industry standards as concerns the labelling of sustainable or green securities and consider, when present, existing policy actions aiming at strengthening sustainable finance.[25] The main benefit of such an approach would be a strong contribution to mainstreaming sustainable finance, by directing an unprecedented amount of financial resources towards specific sectors or activities (the ones considered to foster a more sustainable economy). Eventually, this will also reduce the incidence of sustainability-related risks and in turn would also shield financial sustainability. Nevertheless, some concrete risks can arise from such an approach. On the one side, under-funding dynamics and higher costs of financing could hit sectors not considered as being sustainable, again potentially triggering wide reductions in related assets values. On the other side, and maybe even more importantly, a further widening of the mandate of central banks beyond the traditional primary objective of maintaining price stability might result in a situation in which the effectiveness of the monetary policy could be

---

[25] The labelling of sustainable securities, in particular if needed to drive policy making, is not a straightforward exercise and requires the implementation of a considerable preliminary infrastructure. In this respect, at least two main aspects need throughout consideration. The first concerns the analysis of sectors or activities that can be financed with "sustainable" or "green" funds. The second regards the operational standards (e.g. use of proceeds, management of proceeds, reporting requirements) that need to be followed for labelling a specific security as "sustainable" or "green". For a further dissertation, see Berrou et al. (2019a) and, for the policy activities carried over at the European level in the attempt to mainstream sustainable finance by defining, inter alia, a taxonomy of sustainable activities and correlated labelling standards, https://ec.europa.eu/info/business-economy-euro/banking-and-finance/green-finance_en.

diluted or even endangered.[26] For this reason, throughout prior assessment of the policy and governance implications of such an extension of scope would be necessary.

5.7 Sustainability-Related Risks and Financial Stability: Summary and Preliminary Conclusions

As of today, existing evidence is not sufficient to state strong conclusions on the specific impact of sustainability-related risks on financial stability. Nevertheless, first warnings from international and national institutions have already been launched (e.g. BoE 2015; BIS 2020). Realistically, little probability exists that factors such as climate change, environmental degradation, social inequality, policy and technology shifts will cause in the near term a systemic-wise crisis in the financial system. This notwithstanding, such a possibility is expected to become more concrete in the longer term, in particular if the observed trends linked to climate change and environmental degradation will keep consolidating. In this respect, the harm for financial stability can principally derive from a generalised misinterpretation by financial intermediaries of the magnitude of the challenge ahead. For this reason, understanding the direct and indirect consequences of these new sources of risk on their businesses is an essential preliminary condition to safeguard financial stability. Such an awareness would need to be reflected in the evolution of existing risk-management frameworks and in a recalibration of risk-taking strategies in order to consider the profile over time of sustainability-related risks. On the other side, policymakers are also expected (and in some case have started) to act. In particular, a refinement of the prudential supervisory approaches, by also including instruments able to take into account the features of the different types of sustainability-related risks, is today necessary. Such intervention would be more effective if complementary to wider policy actions to be carried over out in the frameworks given by

---

[26] One can say that in some jurisdictions fostering sustainability should be already considered as a secondary objective of central banks and, as such, can be treated within the existing statutory functions. For example, the main objective of the European Central Banks (ECB) is to "maintain price stability". Nevertheless, "without prejudice to the objective of price stability, the ECB may support the general economic policies in the Union. These may include, inter alia, full employment and balanced economic growth".

the Paris Agreement and the implementation of the Sustainable Development Goals (which, as a by-product, would also allow to reduce the incidence of the sustainability-related risks on the financial risks). In such a context, central banks could assume an unprecedented leading role, as major actors in the supervision of the financial system and potentially able to help mainstreaming sustainable finance. In particular, the integration of sustainability considerations in the execution of the monetary policy would drastically increase the flow of resources directed to finance sustainable activities. However, such possible a new role could carry some relevant drawbacks. This would be principally linked to the need for central banks to consider and integrate a wider policy context and to the possibility to dilute the effectiveness of the monetary policy action in pursuing the primary objective of preserving price stability.

REFERENCES

Allen, W. A., & Wood, G. (2006). Defining and achieving financial stability. *Journal of Financial Stability, 2*(2), 152–172.

Autorité de contrôle prudentiel et de résolution (ACPR). (2019). *French banking groups facing climate change-related risks*. Paris.

Bank for International Settlements (BIS). (2020). *The green swan. Central banking and financial stability in the age of climate change*. Basel.

Bank of England (BoE). (2015, September). *Breaking the tragedy of the horizon— Climate change and stability*. Speech given by Mark Carney, Governor of the Bank of England.

Basel Committee on Banking Supervision (BCBS). (2012, October). *A framework for dealing with domestic systemically important banks*.

Basel Committee on Banking Supervision (BCBS). (2020, April). *Climate-related financial risks: A survey on current initiatives*.

Berrou, R., Ciampoli, N., & Marini, V. (2019a). *Defining green finance: Existing standards and main challenges*. In M. Migliorelli & P. Dessertine (Eds.), *The rise of green finance in Europe. Opportunities and challenges for issuers, investors and marketplaces*. Cham: Palgrave Macmillan.

Berrou, R., Dessertine, P., & Migliorelli, M. (2019b). An overview of green finance. In M. Migliorelli & P. Dessertine (Eds.), *The rise of green finance in Europe. Opportunities and challenges for issuers, investors and marketplaces*. Cham: Palgrave Macmillan.

European Central Bank (ECB). (2019, November). *Financial stability review*. Frankfurt.

European Commission (EC). (2018a). *A Clean Planet for all. A European strategic long-term vision for a prosperous, modern, competitive and climate neutral economy.* COM(2018) 773 final, Brussels.

European Commission (EC). (2015). *Action plan on building a capital markets union.* COM(2015) 468 final, Brussels.

European Commission (EC). (2018b). *Action plan: Financing sustainable growth.* COM(2018) 97 final, Brussels.

European Commission (EC). (2019a). *European financial stability and integration. Review 2019.* Brussels.

European Commission (EC). (2019b). *The European Green Deal.* COM(2019) 640 final, Brussels.

Fell, J., & Schinasi, G. (2005, April). Assessing financial stability: Exploring the boundaries of analysis. *National Institute Economic Review, 192,* 102–117.

International Energy Agency (IAE) and International Renewable Energy Agency (IRENA). (2017). *Perspectives for the energy transition.*

Intergovernmental Panel on Climate Change (IPCC). (2018). *Special Report. Global warning of 1.5 °C.*

Migliorelli, M., & Dessertine, D. (Eds.). (2019). *The rise of green finance in Europe. Opportunities and challenges for issuers, investors and marketplaces.* Cham: Palgrave Macmillan.

National Bank of Belgium (NBB). (2018, June). *Financial stability report 2018.*

Network of Central Banks and Supervisors for Greening the Financial System (NGFS). (2019, April). *A call for action. Climate change as a source of financial risk.*

Prudential Regulation Authority (PRA). (2018, September). *Transition in thinking: The impact of climate change on the UK banking sector.*

Task Force on Climate-related Financial Disclosures (TCFD). (2017). *Recommendations of the Task Force on Climate-related Financial Disclosures.* Basel.

United Nations Environment Programme (UNEP). (2016, September). *Definitions and concepts. Background note* (Inquiry Working Paper 16/13).

University of Cambridge and United Nations Environment Programme—Finance Initiative (Cambridge and UNEPFI). (2014). *Stability and Sustainability in Banking Reform—Are environmental risks missing in Basel III?*

Index

Printed by Printforce, the Netherlands